Contemporary South Asian Studies

Editor-in-Chief

Paulo Casaca, Avenue des Arts 19, South Asia Democratic Forum, Brussels, Belgium

This book series features scientific and scholarly studies focusing on politics, economics and changing societies in South Asia. Utilizing recent theoretical and empirical advances, this series aims at providing a critical and in-depth analysis of contemporary affairs and future developments and challenges in the region. Relevant topics include, but are not limited to, democratization processes, human rights concerns, security issues, terrorism, EU-South Asia relations, regional and economic cooperation and questions related to the use of natural resources. Contemporary South Asian Studies (CSAS) welcomes monographs and edited volumes from a variety of disciplines and approaches, such as political and social sciences, economics and cultural studies, which are accessible to both academics and interested general readers. The series is published on behalf of the South Asian Democratic Forum (Brussels), which is one of the most well-known think tanks in Europe focusing on South Asia.

More information about this series at http://www.springer.com/series/15344

Sumana Bandyopadhyay · Habibullah Magsi ·
Sucharita Sen · Tomaz Ponce Dentinho
Editors

Water Management in South Asia

Socio-economic, Infrastructural,
Environmental and Institutional Aspects

South Asia
Democratic
Forum

Springer

Editors
Sumana Bandyopadhyay
Department of Geography
University of Calcutta
Kolkata, India

Sucharita Sen
SaciWATERs
Secunderabad, Telangana, India

Centre for the Study of Regional
Development, School of Social Sciences
Jawaharlal Nehru University
New Delhi, Delhi, India

Habibullah Magsi
Department of Agricultural Economics
Sindh Agriculture University
Tando Jam, Pakistan

Tomaz Ponce Dentinho
University of Azores
Angra do Heroísmo, Portugal

ISSN 2509-4173   ISSN 2509-4181   (electronic)
Contemporary South Asian Studies
ISBN 978-3-030-35239-4   ISBN 978-3-030-35237-0   (eBook)
https://doi.org/10.1007/978-3-030-35237-0

This Springer imprint is published by the registered company Springer Nature Switzerland AG
The registered company address is: Gewerbestrasse 11, 6330 Cham, Switzerland

Preface

The book, *Water Management in South Asia*, is part of a set of books on regional cooperation in South Asia that has been published by Springer within the activities developed by the South Asian Democratic Forum which is a think tank based in Brussels that aims to contribute to freedom, peace, democracy and development, through the sharing of thoughts on relevant issues such as regional cooperation, international networks, urbanization and sustainability, water management, access to basic services and so on.

The sharing of thoughts on similar issues done by academics, researchers and practitioners from different countries of South Asia and from the world helps to create a common understanding of how to address the issues that, beyond the national and regional borders, are common to many places in South Asia, places that suffer from rural poverty, urban unsustainability, conflicts over natural resources and general difficulties in the provision of adequate public governance. Some of these problems are rooted in policy and science failures in which recognition will nurture improved solutions.

The book *Water Management in South Asia* does that looking at water pollution and depletion, floods and droughts, erosion and sedimentation, but also tries to analyse the economic and environmental impacts of infrastructures and policies and does it for South Asia, eventually one of the more interesting places to know about water issues.

Kolkata, India   Sumana Bandyopadhyay
Tando Jam, Pakistan   Habibullah Magsi
Secunderabad/New Delhi, India   Sucharita Sen
Angra do Heroísmo, Portugal   Tomaz Ponce Dentinho

Acknowledgements The editors acknowledge the suggestions made by Madalena Casaca, Susan Guarda, Stephan Lampe and Elisabete Martins that helped very much in the production of a more understandable text by a potential worldwide community of readers.

Contents

Contributors

Salman Atif Institute of Geographical Information Systems, National University of Sciences and Technology, Islamabad, Pakistan

Nora Babalova South Asia Democratic Forum, Brussels, Belgium

Sumana Bandyopadhyay University of Calcutta, Kolkata, India

Poulomi Banerjee SaciWATERs, Hyderabad, India

Abhijit Chhetri Department of Microbiology, St. Joseph's College, North Point, Darjeeling, India

Ashish Chhetri Department of Geography, St. Joseph's College, North Point, Darjeeling, India

Sandra D'Sa Birla Institute of Technology & Science, Goa, India

Ujjwal Dadhich School of Development Studies, Tata Institute of Social Sciences, Deonar, Mumbai, India

Sayantan Das Department of Geography, Dum Dum Motijheel College, Kolkata, India

Leon M. Hermans Delft University of Technology & IHE Delft, Delft, The Netherlands

Anwar Hussain Department of Economics and Development Studies, University of Swat, Swat, Pakistan

Dhanasree Jayaram Department of Geopolitics and International Relations, Centre for Climate Studies, Manipal Academy of Higher Education (MAHE), Manipal, Karnataka, India

Nabendu Sekhar Kar Department of Geography, Shahid Matangini Hazra Government College for Women, Purba Medinipur, India

Abdullah Abusayed Khan Department of Sociology, Khulna University, Khulna, Bangladesh

G. Mujtaba Khushk Department of Rural Sociology, Sindh Agriculture University, Tandojam, Pakistan

Sk. Mafizul Haque Department of Geography, University of Calcutta, Kolkata, India

Habibullah Magsi Sindh Agriculture University Tandojam, Sindh, Pakistan

Debasis Patnaik Birla Institute of Technology & Science, Goa, India

Tomaz Ponce Dentinho University of Azores, Ponta Delgada, Portugal

Mashal Riaz Institute of Geographical Information Systems, National University of Sciences and Technology, Islamabad, Pakistan

Sucharita Sen Centre for the Study of Regional Development, School of Social Sciences, Jawaharlal Nehru University, New Delhi, India

Abdul Shaban School of Development Studies, Tata Institute of Social Sciences, Deonar, Mumbai, India

M. Javed Sheikh Department of Rural Sociology, Sindh Agriculture University, Tandojam, Pakistan

Taufiq-E-Ahmed Shovo Department of Sociology, Khulna University, Khulna, Bangladesh

Md. Saidur Rashid Sumon Department of Sociology, Jagannath University, Dhaka, Bangladesh

Lakpa Tamang Department of Geography, University of Calcutta, Kolkata, India

Abbreviations

AHP	Analytical Hierarchy Process
APIs	Active pharmaceutical ingredients
BCM	Billion cubic metres
BIS	Bureau of Indian Standards
CBIPMP	Capacity Building for Industrial Pollution Management Project
CD	Community development
CETP	Common effluent treatment plant
CPCB	Central Pollution Control Board
CRED	Centre for Research on the Epidemiology of Disasters
CWD–IMD	Cyclone Warning Division, India Meteorological Department
DDMAs	District Disaster Management Authorities
DEM	Digital elevation model
DFEC	Report of the Damodar Flood Enquiry Committee
DFO	Dartmouth Flood Observatory
DoIW–GoWB	Department of Irrigation and Waterways, Government of West Bengal
DVC	Damodar Valley Corporation
EPS	Ensemble Prediction Systems
FAO	Food and Agriculture Organization
FFD	Flood Forecasting Division
FGD	Focus group discussion
GDP	Gross domestic product
GMP	Ghatal Master Plan
GPS	Global Positioning System
GTV	Ground Truth Verification
HCS	Human-centric sensing
IBIS	Indus Basin Irrigation System
IGB basins	Indus, Ganges and Brahmaputra river basins
IMD	India Meteorological Department
IoT	Internet of Things

IRSA	Indus River System Authority
KP	Kangsabati Project
KPK	Khyber Pakhtun Khwa
LBS	Location-based services
MODIS	Moderate Resolution Imaging Spectroradiometer
NDMA	National Disaster Management Authority
PARC	Pakistan Agricultural Research Council
PCRWR	Pakistan Council of Research in Water Resources
PDMA	Provincial Disaster Management Authorities
PMD	Pakistan Meteorological Department
PS	Participatory sensing
STPs	Sewage treatment plants
TMA	Tehsil Municipal Administrations
UC	Union Councils
WAPCOS	Water and Power Consultancy Services (India)
WAPDA	Water and Power Development Authority
WBDMD	West Bengal Disaster Management Department
WECS	Water and Energy Commission Secretariat
WHO	World Health Organization
WSN	Wireless sensor networks
WWTPs	Wastewater treatment plants

Part I
Conceptual Framework

Chapter 1
Introduction: A Harmonized Approach Towards Water Management in South Asia

Sumana Bandyopadhyay, Habibullah Magsi, Sucharita Sen and Tomaz Ponce Dentinho

1   Introduction

This chapter examines the need for a harmonized framework for the development of guidelines and standards in terms of water management in South Asian regional circumstances as well as to set the examples for regions that have globally similar characteristics. It outlines the proposed framework and the recommendations derived from forthcoming chapters developed by the learned researchers from South Asia and around the globe, to examine the situation. In its simplest form, the framework consists of an iterative cycle, comprising an assessment of water management, with these components being informed by aspects of socio-spatial situations, environmental exposure, and economic positions, from Bangladesh, India, and Pakistan.

South Asia covers about 5.2 million km^2 (2 million mi^2), which is about 3.5% of the world's land surface area. The population of South Asia is about 1.891 billion (UN 2017) or about one-fourth of the world's population, making it both the most populous and the most densely populated geographical region in the world. Indus, Ganges, Brahmaputra, and their tributaries comprise the key watersheds of the subcontinent, crossing different countries and regions. There are wide fluctuations in the river flows

S. Bandyopadhyay (✉)
University of Calcutta, Kolkata, India
e-mail: sumona_bm@yahoo.com

H. Magsi
Sindh Agriculture University Tandojam, Sindh, Pakistan
e-mail: habib_magsi2000@yahoo.com

S. Sen
Centre for the Study of Regional Development, School of Social Sciences, Jawaharlal Nehru University, New Delhi, India
e-mail: ssen@mail.jnu.ac.in

T. Ponce Dentinho
University of Azores, Ponta Delgada, Portugal
e-mail: tomas.lc.dentinho@uac.pt

© Springer Nature Switzerland AG 2020
S. Bandyopadhyay et al. (eds.), *Water Management in South Asia*,
Contemporary South Asian Studies, https://doi.org/10.1007/978-3-030-35237-0_1

during the year. About 90% of the water supplies in the region are used for agriculture and the remaining for households, industry, and other purposes. Canals irrigation is mostly in public sector, and wells/tube wells are generally in private sector.

2 Water Issues

The region possesses only 6.8% of the world's usable water resources, which is lower than its demand for water towards sustainable agriculture and food supplies. Moreover, prolonged dry weather many a times results in droughts and water scarcity in the region. The impact of the drought is of universal nature, affecting vast areas in countries and regions. There were acute shortages of water in regions hit by droughts. The crops damaged, and the pastures dried up. Even access to drinkable water becomes a problem.

The use of excessive mining of water (tube wells) in some areas have lowered the water tables to dangerously low levels. There are massive human and livestock migration from these areas. Thus, the governments of South Asia need to promote policies and devise device management mechanisms for conservation of water resources that optimize use efficiency and bring sustainability of water and life, because water is life.

Transboundary rivers, such as the Ganges, Indus, and Brahmaputra, have defined the geography, history, and culture of South Asia for centuries and are critical to economic growth, food and energy security, and sustainable development within the region (Surie 2015). Nevertheless, over the last few decades, these rivers have come under considerable pressure from industrial development, urbanization, population growth, and environmental pollution. This situation aggravates with poor domestic governance of water resources and increasing variability in rainfall and climate patterns that have made South Asia highly susceptible to floods, droughts, and natural disasters (FAO 2011).

Geopolitics and history of cross-border disputes have meant that transboundary water issues are approach largely from a perspective of national security. This highly securitized approach has severely limited access to hydrological data in the region. South Asian governments, in particular India and Pakistan, treat hydrological data pertaining to transboundary rivers as secret and classified. While some existing bilateral treaties and agreements, such as the Indus Treaty of 1960 (United Nations 1960) or the Ganges Treaty of 1996 (GWT 1996), do contain provisions for bilateral data sharing, actual data-sharing practices are ad hoc, and the range of information shared is quite limited. It is also worth noting that none of the existing treaties provide for the public disclosure of data or information exchange between governments (Surie 2015).

Experts agree that reduced access to freshwater will lead to a cascading set of consequences, including impaired food production, the loss of livelihood security, large-scale migration within and across borders, and increased economic and geopolitical tensions and instabilities. Over time, these effects will have a profound impact on security throughout the region. The scope and scale of these problems demonstrate that hydro-politics is likely to be a crucial variable in South Asian security and will increasingly require a broader understanding of and strengthened institutional capacities for water governance (Koh et al. 2009).

3 What Mechanism Should Be?

For promotion of regional cooperation in the field of water resource in South Asia, the member countries have to respect international laws and principles and agree to consult with one another while utilizing or managing international waters; regional or bilateral cooperation in the field of water resource development is a critical area of consultation and dialogue among nations that share watersheds. A stark reality that will greatly influence water demand in the region is that the population in the South Asian countries is increasing rapidly which is directly proportionate to water demand.

The region will see an increase in the demand for water, thereby creating pressures to increase withdrawals from international rivers. In this context, regional cooperation has become imperative, involving all co-riparian countries in order to manage the increasing demands for water resources. For many South Asian countries, water scarcity is already a daily reality; the current situation is only going to get worse unless more effective governance principles and better management practices sustain adequate actions.

Therefore, there is dire need of four main attitudes regarding water management as follows:

- First, the assumption of riparian rights to water, sediments, and life, detailed by agreement regarding quantity, quality, seasonality, and flood control aiming at the sustainable development of all the river basins.
- Second, the need to avoid free access to ground water, wherein, abuse can lead to the depletion of the water table and thus lead to drastic water shortages.
- Third, agro-ecological researchers have to popularise crop production practices and technologies that are less water intensive.
- Finally, the sharing of data/information between governments, and between governments, civil society, and researchers, might be helpful to think, design, and implement measure for a more sustainable use of water for the peoples and places of South Asia.

4 Contribution of This Book

In South Asia, water and security encompass individual physical safety, livelihoods, health and human welfare, as well as a realization of the cooperative potential between nation states and subnational jurisdictions. The book highlights the significance of water as a basic need, as a source of livelihoods, a potent force behind extreme events and natural disasters. It is also a tool for cooperation towards sustainable management among governments and communities. In particular, beyond this introduction, the book consists of four parts,

- First part gives emphasis on the conceptual framework on how to cope with river basin management issues and politics at regional levels in the South Asian context. Chapter 1 presents a summary of the main river basins of South Asia, and Chap. 2 highlights water politics concerning the management of the Brahmaputra River Basin.
- Second part highlights the trade-offs between availability of water resource and its demand for various uses/sectors; this part articulates the resolution measures regarding the natural calamities that impact the water resource in the rivers and seas of South Asia. Chapter 3 focus the pollution of the an important Indian river flowing through a major urban centre, Godavari River, Chap. 4 looks into the conflicts over groundwater for irrigation, and Chap. 5 pays attention to floods and inundations as a key interceptor of the water-based ecosystems affecting humanity and making communities vulnerable to natural disasters.
- The third part focusses on the water issues at various sectors and regions of South Asia, while it also explains how social capital is playing a crucial role among water communities as well as contributing to sharing experiences regarding water sustainability. Chapter 6 analyses the importance of social capital in the sustainability of water use, Chap. 7 highlights the role of monitoring technologies in disaster prevention, and Chap. 8 focuses on sustainable water management in mountainous areas of the Himalayas.
- Last part of the book mainly consists of institutional coherence (organized proximity) for sustainable use of water at regional levels. This part also recommends that the water conservation management practices should be adopted from other regions as model practices, i.e. mainly those that use the potential of geographical proximity. Chapter 9 studies the economy of water in Pakistan, Chap. 10 analyses the impacts of water logging in Bangladesh, Chap. 11 looks into the prospects of Peacebuilding related to water management in the Indus basin, Chap. 12 highlights water conservation tools in India, and Chap. 13 assesses the impact of urban infrastructures in wetlands.

References

FAO (2011) Climate change, water and food security. FAO Water Reports 36. http://www.fao.org/3/i2096e/i2096e.pdf

GWT (1996) Treaty between the government of the People's Republic of Bangladesh and the government of the Republic of India on sharing of the Ganga/Ganges waters at Farakka. Government of Bangladesh, Dhaka. http://www.hidropolitikakademi.org/treaty-between-the-government-of-the-republic-of-india-and-the-government-of-the-peoples-republic-of-bangladesh-on-sharing-of-the-gangaganges-waters-at-farakka.html

Koh T et al (2009) Asia's next challenge: securing the region's water future. A report by the Leadership Group on Water Security in Asia. https://asiasociety.org/files/pdf/WaterSecurityReport.pdf

Surie MD (2015) South Asia's water crisis: a problem of scarcity amid abundance. March 25, 2015. http://asiafoundation.org/2015/03/25/south-asias-water-crisis-a-problem-of-scarcity-amid-abundance/. Accessed 25 Mar 2015

UN (2017) World population prospects: the 2017 revision. United Nations Department of Economic and Social Affairs, Population Division. Retrieved 10 Sept 2017

United Nations (1960) India, Pakistan and International Bank for reconstruction and development. The Indus Waters Treaty 1960. https://treaties.un.org/doc/Publication/UNTs/Volume%20419/volume-419-I-6032-English.pdf

Chapter 2
Scarce Resource Politics in the Brahmaputra River Basin

Nora Babalova

1 Introduction

> Transboundary water management is a policy arena where domestic politics and international relations intersect. (Xie and Jia 2016, p. 681)

Emerging environmental challenges related to water scarcity, combined with lack of international cooperation regarding shared river flows and unilateral water diversion activities, can trigger instability in various places around the globe, thus increasing the significance of political research regarding Transboundary rivers. Problems linked to international river basins usually relate to conflicting interests among upstream and downstream riparian countries. While upstream countries tend to use water to get more power, downstream ones use power to get more water (Warner 2004).

Transboundary water affairs can be approached from various angles. Two most influential theories applied to international relations, (Neo)realism and (Neo)liberalism, highlight on the one hand the aspects of international interdependence and rule-based behaviour, and on the other hand, the 'power factor' emphasising the state's own interest—which is not to be subordinated to the interests of other states. Besides mainstream theories on international relations, international waters can be studied from the perspective of political ecology and political geography, as well as from the point of view of other relevant disciplines such as geography, hydrology, ecology, and law. Some approaches even incorporate disciplines such as anthropology and sociology (Price and Mittra 2016).

Conventional analysis, however, tends to overlook the role that power asymmetry plays in the context of international rivers; the conceptual framework of hydro-hegemony fills this gap by providing an 'analytical paradigm useful for examining the options of powerful or *hegemonised* riparians and how they might move away

N. Babalova (✉)
South Asia Democratic Forum, Brussels, Belgium

© Springer Nature Switzerland AG 2020
S. Bandyopadhyay et al. (eds.), *Water Management in South Asia*,
Contemporary South Asian Studies, https://doi.org/10.1007/978-3-030-35237-0_2

9

from domination towards cooperation' (Zeitoun and Warner 2006, p. 435). Following Lowi's (1993) hegemonic stability theory, the chapter assumes that the 'state which is the furthest upstream and hence, in the most favourable geographic position, will have no obvious incentive to cooperate', allowing the hegemon to 'utilize as much of the water as it chooses unilaterally, irrespective of downstream needs' (p. 10). A 'hegemonised riparian' must have the following three elements in order to be defined as a hydro-hegemon: ability to dominate other countries (geographic, political, economic, and military superiority); incentives to dominate other countries; and the potential to benefit from exceeding water rights, harming the interests of downstream countries and creating an unfair water distribution (Wang 2015).

The present chapter applies the hydro-hegemony framework to the relatively under-studied Brahmaputra River basin, an endeavour we judge most useful given the obvious (a) power asymmetry that dictates how the river's resources are allocated and (b) weak international institutional context. The Brahmaputra case is studied from a political perspective wherein China is perceived as the local hydro-hegemon.

Various scholars predict that competition over the Brahmaputra's resources can lead to serious discord between the densely populated and water-stressed riparians. The major issue relates to the fact that without a proper basin-wide agreement, an upstream country can easily limit the water flow and thus cause unfair water distribution (Chellaney 2009, 2011, 2013; Economy and Levi 2014; Ramachandran 2015; Samaranayake et al. 2016; Wuthnow 2016). Most agreements that relate to multilateral river basins (shared by three or more states) such as the Brahmaputra are of a bilateral nature and thus lack comprehensiveness (Zawahri and Mitchell 2011). Furthermore, China is one of three countries that voted against the 1997 UN Convention on the Law of the Non-Navigational Uses of International Watercourses, an attitude revealing the country's unwillingness to engage in any form of international cooperation (Jayaram 2012). Besides the power asymmetry factor and China's hegemonic position, the creation of an institutionalised form of cooperation is hindered by unresolved border disputes between two major riparians, China and India. There are two main disputed border areas between China and India: the western border of Aksai Chin bordering Kashmir and the territory of Arunachal Pradesh in Northeast India. The Brahmaputra River flows through the latter. Although China and India signed several bilateral agreements so as to ease political tensions, the danger of water-related conflicts in the basin may still seem present.

However, the much-feared but non-existent 'water wars' are yet to take place. It is believed that these conflicts never actually lead to war because of the imbalance of power between the riparian states rather than because of a perceived cooperation (Zeitoun and Warner 2006). The latter is consistent with Frey's (1993) power analytical framework, stating that the 'least stable [situation] is when the downstream nation is most powerful and has most interest in water but the upstream nations also have considerable interest' (p. 62). Yet, the opposite is true for the distribution of power asymmetries in the studied Brahmaputra basin, wherein the downstream riparian (Bangladesh) is the weakest concerned political actor and the upstream riparian (China) is the strongest. Here, the chapter's argument is consistent with scholars emphasising that a 'significant factor preventing war over water is that the actions of non-hegemonic states usually comply with the order preferred by the hegemon, whose superior power position effectively discourages any violent resistance against the order'. The hydro-hegemon may, therefore, decide to enforce either 'negative' (oppressive domination) or 'positive' (enlightened leadership) forms of dominant hydro-hegemony, the latter being an option whereby all basin countries benefit (Zeitoun and Warner 2006, p. 437, 439).

Following the theoretical rationale, the chapter provides the background necessary for understanding the state of water management and (non) cooperation between riparian countries. It then investigates the major hindrances to effective basin-wide collaboration, in an attempt to seek a realistic path towards a harmonious 'water order'. The chapter concludes with a concise discussion within the aforementioned theoretical framework and explores how riparian countries could benefit from greater cooperation, even in the absence of a formal water management regime.

2   The Brahmaputra Case

The Tibetan Plateau is the Asian continent's water tank and the source of all major rivers. Rivers in South Asia are usually shared by two or more countries. Originating in the Himalayas, the Brahmaputra River passes through Southwest China (Tibet), two Indian states (Arunachal Pradesh and Assam), and Bangladesh. Locally, the river is known as Yarlung Tsangpo in Tibet, Jamuna in Bangladesh, and Brahmaputra in India (meaning the 'Son of Brahma' in Sanskrit). For consistency purposes, the chapter uses the latter term so as to identify the river throughout the whole basin.

Due to the rugged and high-altitude terrain of the Brahmaputra basin and the risk of water-related conflicts with neighbouring countries, it was long thought impossible to access the enormous energy resources of the world's highest-altitude river (Watts 2010a, b). The Brahmaputra does possesses fewer dams than the other two major river basins in South Asia, the Indus and Ganges (Samaranayake et al. 2016). However, in response to both China's and India's rising energy needs, growing populations, and rapid urbanisation, the number of dams along the Brahmaputra River started to increase dramatically, as evidenced by the ongoing dam constructions in both Tibet and Northeast India (especially in Arunachal Pradesh).

Holding the highest hydropower potential in the world, China perceives the rivers of Tibet as a new area of hydropower development (Buckley 2015). The area is thus considered an essential element of Beijing's 'Open Up the West' campaign intended to encourage economic development in Western China, which was launched in 2000 (Samaranayake et al. 2016). As the waterways of mainland China were already packed with dams, China began constructing a new series of dams on sections upstream from the Brahmaputra Canyon, thus ending the Brahmaputra's status as the world's last undammed river (Chellaney 2013; Buckley 2015).

India and Bangladesh experienced the negative impacts of China's dams. Both countries depend on the Brahmaputra's resources—mainly for rice and agricultural crops. The fish caught during the flood season is the main source of food for the majority of rural population living in the basin. Flowing through two Indian states of Arunachal Pradesh and Assam, the Brahmaputra accounts for about thirty per cent of India's water runoff, constituting a major source of the country's potential hydropower (Chaturvedi 2013). In Bangladesh, up to seventy per cent of population resides in the Brahmaputra basin—compared to only one per cent of China's and three per cent of India's population (Samaranayake et al. 2016). Given these imbalances, the threat perception is much larger on the side of India and Bangladesh. As the source of the Brahmaputra River lies in China, the river flow can be unilaterally limited or regulated at any time, for example, through dam constructions or water diversion projects.

3   China's Activities on the Tibetan Plateau: Expansion of Renewable Power in the Himalayas

As the largest global producer of hydropower, China boasts a greater number of large dams on its territory than the rest of the world combined: if dams of all sizes are counted, China has around 90,000 (Chellaney 2013). The country has been developing various water transfer projects since 2000 (Xie and Jia 2016, p. 682). The broadly discussed South-to-North Water Diversion Project, which transfers water from the Yangtze River to the water-stressed cities of Beijing and Tianjin, dramatically changed the face of China's countryside and displaced great masses of people (Kaiman 2014; Ghassemi and White 2011). Holding a record number of the world's worst dam-related disasters, the human costs of China's water projects have always been high (Lovell 2016). The world's largest hydropower project, the Three Gorges dam on the Yangtze River in Central China, for instance, displaced around 1.7 million people and submerged 13 cities, 140 towns, and 1350 villages (Chellaney 2013; Lewis 2013). The dam was also blamed for worsening water pollution levels due to river depletion and sedimentation (Gleick 2009).

Recently, China has shifted its dam-building focus from internal rivers in the Han heartland to international rivers that originate in remote and ecologically sensitive parts of the Tibetan Plateau—the source of all major rivers in the region (Chellaney 2013). The livelihoods of almost two billion people residing in South and Southeast Asia would be affected if the water resources of the Tibetan Plateau are altered by China's activities such as dam constructions, cutting down forests, and mining (Jayaram 2012). Yet, this is exactly what Beijing is undertaking: constructing gigantic dams in remote areas of Tibet which threaten both local biological and ethnic diversity and deeply affect downstream flows. As part of an even larger water diversion plan, dams on the upper reaches of the Brahmaputra are intended to divert water so as to feed China's 'desperate thirst for clean water' (Buckley 2014, p. 13). These matters, however, are not discussed in public so as to avoid bitter reactions from both the international community and the downstream riparians India and Bangladesh

(Chellaney 2009). When protests against hydropower projects occur, they are either prohibited or violently suppressed (Buckley 2014, 2015).

The Zangmu Dam, China's first hydroelectric dam on the Brahmaputra River in Tibet, which became fully operational in October 2015, will be soon followed by other gigantic dam projects (Samaranayake et al. 2016). One announced project is a dam on the Brahmaputra at the Great Bend, in a county perceived as a sacred region by many Tibetans, and where the river takes a U-turn before entering Arunachal Pradesh in India. If this project is undertaken, this would become the world's largest dam and hydropower station (Watts 2010a, b; Lewis 2013). Opponents to the project warn about the seismically active and fragile character of the area: while the dam would probably save a sizeable amount of carbon dioxide, it could lead to conflict over downstream water supplies, as blocking the upper reaches of the Brahmaputra would devastate the fragile ecosystem of the Tibetan Plateau and withhold the river's sediments from the fertile floodplains in north-eastern India and Bangladesh (Watts 2010a, b; Lewis 2013). Furthermore, the local population will hardly benefit from these monumental dams, given that the generated energy will be transferred to the power-hungry and water-stressed cities of Beijing and Tianjin farther east while many Tibetans will become deprived of their land. Considering the political and security aspects involved, it is equally relevant that the dam is supposed to be located in proximity to the McMahon Line,[1] a disputed and militarised boundary between China and India.

4   Competition Over Brahmaputra's Scarce Resources: The Plight of Non-cooperation

Human welfare cannot be measured only through economic growth and development, but also should consider the current ecological limits of the planet, allow for fair distribution, and promote an efficient allocation of resources. (Salgado 2015)

The drive to construct dams which affect river flows has long the subject of much controversy (Phillips et al. 2006). Both China and India seem passionate about gigantic hydropower projects, building more dams than the rest of the world combined. Major reasons for the increased production of large-scale dams in the Brahmaputra basin include urbanisation, rising energy needs, and poverty reduction benefits (Welford 2010). The basin countries are 'currently undergoing rapid industrialisation

---

[1] The McMahon Line constitutes a border between India and China, extending from east Bhutan to Myanmar. Historically, the Line is based on a British colonial claim, drawn as the frontier between British India and Tibet at the Shimla Conference in 1914. In 1949, however, the newly formed People's Republic of China refused to accept the McMahon Line. India, on the other hand, regarded the McMahon Line as a permanent international border. This discord became a major cause of tension between China and India in the 1950s and one of the main reasons for the Sino-Indian war of 1962. Today, the border is known as the Line of Actual Control that separates Indian-ruled land from Chinese-controlled territory (Mitra et al. 2006; Florcruz 2013).

while facing the pressing needs of improving their water management capacities and fighting climate change'; it is thus cause for concern how these countries will cope with the 'challenge of realizing sustainable water resource management' (Xie and Jia 2016, p. 678).

India has been obsessed with large dams since the 1960s, starting with the construction of the Bhakra Dam in northern India. This dam became the symbol of India's Green Revolution.[2] It was hailed by the then Prime Minister Jawaharlal Nehru as a 'Temple of Modern India',—although he later came to regard it as the 'disease of gigantism', as badly managed irrigation schemes resulted in waterlogged, saline soils, and diminishing harvests (Bosshard 2015). One of India's most controversial dam project, however, is the Sardar Sarovar Dam[3] situated on the Narmada River, due to which over 250.000 citizens (mostly indigenous peoples), were displaced during the 1980s (Mitra et al. 2006; Bosshard 2015).

With regard to the Brahmaputra River, the river and its tributaries in Arunachal Pradesh are considered a new hotspot for India's dam building and future powerhouse of the country. Being home to numerous indigenous tribal groups, Arunachal Pradesh is the most geographically isolated state in India, which attracts the attention of dam builders (Overdorf 2012). A total of 168 massive dams were proposed in Northeast India, the majority of which are to be located in Arunachal Pradesh, including the largest of them all, the Lower Subansiri Hydroelectric project (Yong 2014; Overdorf 2012).

Dams can play a vital role in enhancing economic growth and social development, but they can also have disastrous consequences as they impoverish masses of people (Welford 2010) and deeply affect downstream ecosystems, river-dependent populations, and water flows (Yong 2014). The damage that dams causes to river ecosystems can be immense, 'turning free-flowing waterways into lifeless lakes, killing plants and trees, blocking fish migration and breeding, driving species to extinction, and devastating established patterns of human life' (Lewis 2013). In Arunachal Pradesh, for instance, it is expected that over 50,000 acres of forest will be submerged by the proposed hydroelectric projects (Overdorf 2012). International concerns about the environmental and social impacts of large-scale dams have been increasing since the beginning of the twenty-first century. Both urbanisation and states' preference for technocratic solutions for energy production are responsible for the rapid growth of gigantic hydropower projects in South Asia. Large-scale hydropower projects in areas

---

[2] In the Indian context, the Green Revolution refers to the period from 1967 to 1978, during which an economic and agricultural development strategy introduced by Western aid organisations was applied to make India self-sufficient in food grains (Mitra et al. 2006).

[3] The Sardar Sarovar Dam was first envisaged in the 1940s by Jawaharlal Nehru to control the immense Narmada River system. The World Bank financed the dam construction even though the project did not comply with the government's conditional environmental clearance. In the mid-1980s, the Narmada Bachao Andolan (Save the Narmada Movement), a coalition of social movements and NGOs, created strong international public pressure to stop the dam. As a result, the World Bank had to withdraw from the project in 1994 after an independent review found systematic violations of its social and environmental policies (See Mitra et al. 2006; Bosshard 2008, 2015).

prone to natural disasters may, however, result in negative social and environmental impacts that can spur new migration or refugee flows.

Protests against gigantic dams occur against both domestic and upper riparian's projects; however, they tend to be little reported by the media (Sharma 2012). Several anti-dam organisations raising voices against the completion of China's hydropower project on the upper reaches of the Brahmaputra in Tibet took place in 2014 in Assam; some activists even threatened to lay a siege on the Chinese embassy in New Delhi (Lewis 2013; Karmakar 2014). On a domestic level, demonstrations recently occurred in Tawang,[4] a small town in Arunachal Pradesh, during which at least two protesters were killed by the police (The Third Pole 2016). In November 2016, during a meeting on 'Policy Dialogue for Governance of the Brahmaputra River' in Arunachal Pradesh, a group of anti-dam leaders confronted government officials, clarifying that they protest neither against small dams nor against dams as such, but that they cannot support the construction of gigantic dams that threaten their livelihoods, for such large dams will inevitably flood fertile agricultural land, destroy local fauna and flora, and cause displacements (Duarah 2017). The prevailing lack of trust is in part caused by the fact that many people who lost their land due to development projects have not received proper compensation (Chakma and Shimray 2016).

Bangladesh, as the lowest riparian, has been the strongest promoter of multilateral cooperation on the Brahmaputra River. The Brahmaputra is Bangladesh's largest water system, providing approximately sixty-five per cent of the country's river water per year. Moreover, up to seventy per cent of Bangladesh's population resides in the basin—again, compared to only one per cent of China's and three per cent of India's populations (Samaranayake et al. 2016). Bangladesh faces various natural challenges such as floods, riverbank erosion, or diminished water flows; however, climate change projections forecast even more floods and stronger cyclones in near future (McVeigh 2017).

In order to avoid potential negative scenarios, decision makers need to explore and assess all viable alternatives and related impacts before they take a final decision. It is particularly crucial to assess whether a large dam would fulfil (a) the needs of the region and (b) the environmental requirements of the (disaster-prone) Brahmaputra basin. Such alternatives might include renewable energy collected from sunlight, wind, and tides (Bosshard 2015) or small-scale hydroelectric schemes (Salgado 2015). Initiatives like the World Commission on Dams (WCD), established in response to growing opposition to large dam projects in 1998, aim to review the development effectiveness of dams and develop standards and guidelines for future dams and dam building activities. They also aim to highlight the importance of recognising rights, addressing risks, and safeguarding the entitlements of all groups of affected

---

[4]Tawang is part of the unresolved border issues between India and China: to the north it borders Tibet and to the south-west it lies next to Bhutan (Mitra et al. 2006). This region was occupied by Chinese troops in 1962 and, together with the rest of Arunachal Pradesh, China continues to claim the entire area, considering it part of South Tibet.

people (including vulnerable groups and tribal peoples), as well as exploring possible alternatives and monitoring existing dams.

Most importantly, however, cooperation among basin countries needs to be enhanced, either through a regime change or through incentives. China should provide access to more detailed information regarding its dam construction plans on the Brahmaputra so as to improve trust. India should enhance hydrological data sharing between the centre and its northeast governments and inform local governments about all dam construction projects plans in the region. Bangladesh, as the third main riparian, needs to seek assistance from the international community so as to conduct evidence-based assessments of human security impacts in the Brahmaputra basin area (Samaranayake et al. 2016, p. 90).

5 Conclusion: The Potential for Greater Cooperation Within the Context of Power Asymmetries?

River water sharing can be a source of both cooperation and conflict. China's activities on the Tibetan Plateau—the source of all major rivers in South Asia—have a significant impact on downstream neighbours. China's actions in the Himalayas focus on controlling natural environments in pursuit of national goals such as diverting water for purposes of electricity generation, irrigation, mining, and other economic activities (Jayaram 2012). It is perhaps time to consider the wider impacts of such activities.

Going back to the argument presented in the introduction, China possesses all three elements of hydro-hegemony (Wang 2015): both the ability and incentives to dominate other riparians and harm their interests by overpowering water resources. Given that China has not signed a single treaty with any of its neighbours on river water sharing, it is able to exploit water as a tool to persuade lower riparian neighbours to support China's interests. It seems that there is little or no desire from China to share Brahmaputra waters; indeed the country plans to build the world's largest dam and hydropower station at the Great Bend of the Brahmaputra River. The dam represents the ultimate hope for China's exploitation of water resources because it could generate energy equivalent to all the oil and gas available in the South China sea (Watts 2010a, b). In order to minimise the risk of water-related conflicts, China and India have agreed to share information about plans on the Brahmaputra River as well as hydrological information through a Memorandum of Understanding; however, the 'absence of a treaty makes it next to impossible for India to verify China's claims' (Jayaram 2012; Watts 2010a, b). Although some form of cooperation exists, it is largely consolidated in the stronger riparian's favour.

The present chapter showed how domestic politics (i.e., China's water policies and its activities on the Tibetan Plateau) intersect with international relations. As far as the idealistic idea of a basin-wide water regime is concerned, one can conclude that an institutionalised form of cooperation in the Brahmaputra basin remains rather utopian as long as China, who enjoys hegemonic status both geographically and militarily,[5] is neither willing nor forced to accept one (Lowi 1993, p. 203). Hence, the distribution of power and geography dictates the basin regime.

Interactions between China and India over the Brahmaputra's resources could be interpreted as lying somewhere between 'genuine cooperation and cut-throat competition' (Zeitoun and Warner 2006, p. 443), with each riparian trying to maximise its own objectives. The territorial dispute over Indian-administered Arunachal Pradesh and the general lack of trust with regard to the ongoing dam constructions in both Tibet and Northeast India, present major hindrances to larger cooperation between the two countries. Besides, there exist visible tensions regarding China's strengthening relations with countries in the Indian Ocean region on one side and India's ambitions to explore oil and gas in the South China Sea on the other. These obstacles translate into low levels of regional cooperation, making the Brahmaputra basin a textbook example of the least-coordinated river systems in the world (Xie and Jia 2016).

Nonetheless, despite the rather turbulent past and the 2017 Doklam standoff, China and India managed to develop strong economic relations. In April 2018, the two held an informal summit in Wuhan which, according to some, signalled that economic pragmatism and the prospect of trade might overcome (or at least overshadow) long-standing disputes. Therefore, following Lowi's (1993) rationale, the prospect of stronger regional trade ties might incentivise China to manage the Brahmaputra's resources in a more responsible way. Trade could thus become *the* driver for closer cooperation between the basin countries, even in the absence of a comprehensive water regime: mutually beneficial, 'shared interest' water projects, for instance, can serve as an effective incentive for cooperation which can lead to more stable water relations (Zeitoun and Warner 2006, p. 447). To turn this into a reality, however, the major riparians—China and India—need to address the immense trust deficit, which remains a substantial challenge.

References

Bosshard P (2008) New independent review documents failure of Narmada Dam. International Rivers. Retrieved from https://www.internationalrivers.org/blogs/227/new-independent-review-documents-failure-of-narmada-dam

Bosshard P (2015) 12 dams that changed the world. The Guardian. Retrieved from https://www.theguardian.com/environment/blog/2015/jan/12/12-dams-that-changed-the-world-hoover-sardar-sarovar-three-gorges

Brahmaputra Focus Area Strategy: 2013–2017 (2017) South Asia water initiative. Retrieved from https://www.southasiawaterinitiative.org/category/brahmaputra-basin/

Buckley M (2014) Meltdown in Tibet: China's reckless destruction of ecosystems from the highlands of Tibet to the Deltas of Asia, Illustrated edn. St. Martin's Press, New York

Buckley M (2015) The price of Damming Tibet's Rivers. The New York Times. Retrieved from https://www.nytimes.com/2015/03/31/opinion/the-price-of-damming-tibets-rivers.html

---

[5]In terms of military expenditures; land, air, and navy forces (not in terms of manpower).

Chakma T, Shimray GA (2016) South Asia: India. The Indigenous World 2016. International Work Group for Indigenous Affairs, Copenhagen, pp 328–341

Chaturvedi MC (2013) Ganga-Brahmaputra-Meghna waters: advances in development and management. CRC Press, Boca Raton

Chellaney B (2009) The Sino-Indian water divide. Project Syndicate. Retrieved from https://www.project-syndicate.org/commentary/the-sino-indian-water-divide?barrier=accesspaylog

Chellaney B (2011) Water: Asia's new battleground. Georgetown University Press, Washington, DC

Chellaney B (2013) Water, peace, and war: confronting the global water crisis. Rowman & Littlefield Inc, Lanham

Duarah CK (2017) Protests against dams on the Siang continue in Arunachal Pradesh. The Quint. Retrieved from https://www.thequint.com/news/environment/protests-against-dams-on-the-siang-continue-in-arunachal-pradesh

Economy E, Levi MA (2014) By all means necessary: how China's resource quest is changing the world. Oxford University Press, Oxford

Florcruz M (2013) Line of actual control: China and India again squabbling over disputed Himalayan Border. International Business Times. Retrieved from https://www.ibtimes.com/line-actual-control-china-india-again-squabbling-over-disputed-himalayan-border-1236401

Frey FW (1993) The political context of conflict and cooperation over international river basins. Water International. 18(1):54–68

Ghassemi F, White I (2011) Inter-basin water transfer: case studies from Australia, United States, Canada, China, and India. Cambridge University Press, Cambridge, UK

Gleick PH (2009) Three Gorges Dam project, Yangtze River, China (Water Brief 3). The World's Water 2008–2009. Island Press, Washington, DC, pp 139–150

Jayaram D (2012) Environmental change and ripples for water security in Southern Asia. NTS Alert, RSIS Centre for Non-Traditional Security (NTS) Studies. Retrieved from http://www3.ntu.edu.sg/rsis/nts/HTML-Newsletter/Alert/pdf/NTS_Alert_jul_1201.pdf

Kaiman J (2014) China's water diversion project starts to flow to Beijing. The Guardian. Retrieved from https://www.theguardian.com/world/2014/dec/12/china-water-diversion-project-beijing-displaced-farmers

Karmakar R (2014) Assam protests China dam on upper Brahmaputra. Hindustan Times. Retrieved from https://www.hindustantimes.com/india/assam-protests-china-dam-on-upper-brahmaputra/story-Ar1t9f1ciAxUn14Z7ZOxKL.html

Lewis C (2013) China's Great Dam Boom: a major assault on its Rivers. The Yale School of Forestry & Environmental Studies. Retrieved from https://e360.yale.edu/features/chinas_great_dam_boom_an_assault_on_its_river_systems

Lovell S (2016) China's mega hydro-projects are coupled with devastating impacts. Pulitzer Centre. Retrieved from http://pulitzercenter.org/reporting/chinas-mega-hydro-projects-are-coupled-devastating-impacts

Lowi M (1993) Water and power—the politics of a scarce resource in the Jordan River Basin. Cambridge University Press, Cambridge, USA

McVeigh K (2017) Bangladesh struggles to turn the tide on climate change as sea levels rise. The Guardian. Retrieved from https://www.theguardian.com/global-development/2017/jan/20/bangladesh-struggles-turn-tide-climate-change-sea-levels-rise-coxs-bazar

Mitra SK, Wolf SO, Schöttli J (2006) A political and economic dictionary of South Asia: an essential guide to the politics and economics of South Asia, 1st edn. Routledge, London

Overdorf J (2012) Part 1: how many dams can one state hold? Public Radio International. Retrieved from https://www.pri.org/stories/2012-04-23/part-1-how-many-dams-can-one-state-hold

Phillips DJH, Daoudy M, Öjendal J, Turton A, McCaffrey S (2006) Trans-boundary water cooperation as a tool for conflict prevention and for broader benefit-sharing. Ministry for Foreign Affairs, Stockholm, Sweden

Price G, Mittra S (2016) Water, ecosystems and energy in South Asia: making cross-border collaboration work; the research paper of the Chatham House. The Royal Institute of International Affairs, London

Private dam builders back out of Brahmaputra dams (2016, February). The Third Pole. Retrieved December 15, 2016, from https://www.thethirdpole.net/en/2016/02/25/private-dam-builders-back-out-of-brahmaputra-dams/

Ramachandran S (2015) Water wars: China, India and the Great Dam Rush. The Diplomat. Retrieved from https://thediplomat.com/2015/04/water-wars-china-india-and-the-great-dam-rush/

Samaranayake N, Limaye S, Wuthnow J (2016) Water resource competition in the Brahmaputra River Basin: China, India, and Bangladesh. CNA. Retrieved from https://www.cna.org/cna_files/pdf/IRM-2016-U-013097.pdf

Salgado N (2015) Many small vs a few big: alternatives to Megadams in the Chilean Patagonia. Solutions J 6(1):34–39. Retrieved from https://www.thesolutionsjournal.com/article/many-small-vs-a-few-big-alternatives-to-megadams-in-the-chilean-patagonia/

Sharma T (2012) Fighting India's mega-dams. Chinadialogue. Retrieved from https://www.chinadialogue.net/article/show/single/en/4799

Wang Z (2015) Hydro-hegemony, security order and institution construction: on international rivers political complex. Capacity4dev. Retrieved from https://europa.eu/capacity4dev/hhlawaid/minisite/hydro-hegemony-security-order-and-institution-construction-international-rivers-political-c

Warner J (2004) Mind the GAP—Working with Buzan: the Illisu Dam as a security Issue. SOAS Water Issues Study Group, School of Oriental and African Studies/King's College, London (Occasional Paper 67)

Watts J (2010a) Chinese engineers propose world's biggest hydro-electric project in Tibet. The Guardian. Retrieved from https://www.theguardian.com/environment/2010/may/24/chinese-hydroengineers-propose-tibet-dam

Watts J (2010b) When a billion Chinese jump: how China will save mankind—or destroy it. Scribner, New York

Welford R (2010) The future of dam building. CSR Asia. Retrieved from http://www.csr-asia.com/weekly_news_detail.php?id=12084

Wuthnow J (2016) Water war: this river could sink China-India relations. The National Interest. Retrieved from https://nationalinterest.org/feature/water-war-river-could-sink-china-india-relations-15829

Xie L, Jia S (2016) Diplomatic water cooperation: The case of Sino-India dispute over Brahmaputra. Int Environ Agreements Polit Law Econ 17(5):677–694

Yong Y (2014) World's largest hydropower project planned for Tibetan Plateau. Chinadialogue. Retrieved from https://www.chinadialogue.net/article/show/single/en/6781-World-s-largest-hydropower-project-planned-for-Tibetan-Plateau

Zawahri NA, Mitchell SM (2011) Fragmented governance of international rivers: negotiating bilateral versus multilateral treaties. Int Stud Q 55:835–858

Zeitoun M, Warner J (2006) Hydro-hegemony—a framework for analysis of trans-boundary water conflicts. Water Policy 8(5):435–460

Part II
Water Trade-offs Between Nature and Mankind

Chapter 3
The Impact of the Pharmaceutical Industry of Hyderabad in the Pollution of the Godavari River

Sandra D'Sa and Debasis Patnaik

1 Introduction

There are an estimated 12,000 human and 2500 veterinary pharmaceutical products worldwide. These substances—as well as their degraded and externality-related remains—persist in the environment for long periods of time after use (Sayadi et al. 2010).

The presence of pharmaceutical residues in the environment and related health and environmental implications is a concern of increasingly global proportions. Several publications have brought this issue to the attention of the scientific community and policy makers (Hernando et al. 2006; Balakrishna et al. 2017; Shalini et al. 2010; Larsson 2014).

Pharmaceuticals present in effluent wastewaters of drug-manufacturing plants are considered contaminants or pollutants. There is a continuous release of pharmaceuticals along with their metabolites into the environment (Daughton and Ternes 1999).

These pollutants are also termed 'pseudo-persistent compounds' because of their continuous leak into the environment as a result of improper disposal by manufacturing companies (Ankley et al. 2007).

Pharmaceuticals impregnated into the environment—most specifically into water flows—come from both indirect and direct sources. Three sources contribute to this infusion: humans themselves, agricultural activities and manufacturing processes by pharmaceutical companies (Gadipelly et al. 2014).

Pharmaceuticals expelled into the environment include excretion of unreacted compounds and metabolites by human (Tischler et al. 2013) or animal users, the

S. D'Sa (✉) · D. Patnaik
Birla Institute of Technology & Science, Goa, India
e-mail: p2013104@goa.bits-pilani.ac.in

D. Patnaik
e-mail: marikesh@goa.bits-pilani.ac.in

© Springer Nature Switzerland AG 2020
S. Bandyopadhyay et al. (eds.), *Water Management in South Asia*,
Contemporary South Asian Studies, https://doi.org/10.1007/978-3-030-35237-0_3

direct disposal of leftover drugs (Caldwell 2016), and the effluents discharged from manufacturing facilities. Release pathways include domestic sewage systems, discharges from wastewater treatment plants (WWTPs), landfills, and the direct release of human, industrial and animal wastes to both surface and ground water systems.

Pharmaceutical compounds consumed as medicine by humans find their way through human waste into the sewage system. The sewage treatment plants (STPs) do not possess the technology necessary to eliminate pharmaceutical compounds from treated water. This treated water is in turn released into waterbodies at large—contributing to the presence of pharmaceuticals in aquatic environments. Pesticides and other related compounds used for the purpose of preventing destruction of crops have a pathway through agricultural run-off during the irrigation process or during rain periods. Both human and agricultural-related sources are considered indirect sources. The direct source of water contamination is that implemented by pharmaceuticals manufacturers themselves through an often inadequate, inefficient and ineffective wastewater management (Fig. 1).

Pharma residues, even in low concentrations levels of mg/l or μg/l, can have toxic effects. This is the case regarding steroids (Caldwell et al. 2008), which are known to cause endocrine disruption, and in particular feminisation of fish (Synder et al. 2008). Antibiotics in surface water cause antimicrobial resistance (AMR) (Nordea 2016). AMR today has been recognised as a crisis of global proportions, as drug residues have been found in a variety of freshwater environments such as rivers, lakes and even ground waters (Fick et al. 2009).

The wastewater treatment plant located at Patancheru, Patancheru Enviro Tech Ltd. (PETL) receives approximately 1500 m³ of wastewater per day, mainly from approximately 90 bulk drug manufacturers. It discharges its treated effluent into the Isakavagu stream, which flowed into the Nakkavagu and from there into the Manjira tributaries of the Godavari River. Investigation by Larsson et al. (2007) showed that

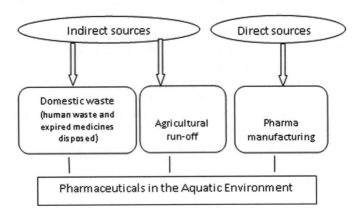

Fig. 1 Entry sources of pharmaceuticals into the aquatic environment. *Source* Adapted and modified from Caldwell et al. (2016) and Gadipelly et al. (2014)

extraordinarily high levels (mg/L) of several drugs were present in this treated effluent and thus transported into the Godavari River.

Seven fluoroquinolones—ciprofloxacin, lomefloxacin, ofloxacin, norfloxacin, enrofloxacin, pefloxacin and difloxacin—have been much investigated by researchers. Ciprofloxacin was chosen to be presented in Table 1 as it is the most common contaminant—and one showing much higher values than other fluoroquinolones.

Larsson et al. (2007) found that Active Pharma Ingredients (APIs) were also discharged during production. This conclusion was reached by analysing wastewater characteristics present in a common effluent treatment plant (CETP) at Patencheru—tasked with treating wastewaters from 90 bulk drug manufacturers in Patencheru Industrial Estate in Hyderabad. The samples contained the highest levels of pharmaceuticals reported in any effluent. It was found that the quantified value of API ciprofloxacin in Hyderabad's wastewaters was equivalent to the five-day dosage requirement for an entire town in Sweden of nine million people. The clarified treated effluents were discharged into the Isakavagu stream which eventually reaches the Godavari River through the Nakkavagu and Manjira tributaries flowing 5 km from the Patancheru Industrial Estate.

This chapter addresses the impact of the direct source of water pollution by Hyderabad's pharmaceutical industry on the Godavari watershed.

Table 1 Levels of ciprofloxacin in ng/L (nanogram per litre) (antibiotic in class of fluoroquinolones) in surface waters in various locations in Hyderabad as compared to a few other countries

Country	Type of water	Sampling site	Concentration in ng/L	Reference
India, Hyderabad	Surface water	Undiluted effluent from Patancheru Effluent Treatment Limited (PETL)	14,000,000	Fick et al (2009)
India, Hyderabad	CETP effluent	Patancheru	28,000,000–31,000,000	Larsson et al. (2007)
India, Hyderabad	Surface water	Musi River located upstream of Amberpet WWTP	7447.0–5,015,690.0	Gothwal and Shashidhar (2016)
China, Spain, Italy, USA	Surface water	Lake Baiyangdian, Jarama River, Po River, Tennessee Upper River Basin	9.4, 112.75, 8.8, (4.7–54.2)	Gothwal and Shashidhar (2016)

2   Objectives

This chapter provides an overview of the pollution impact by the pharmaceutical industry in Telangana—more specifically in the areas in and around Hyderabad—on the Godavari River.

This chapter has the following aims:

1. To review existing knowledge regarding the social, economic and environmental impacts of pharmaceutical pollution in the area.
2. To highlight technological an institutional/management-related solutions intended to reduce effluent discharges from the pharmaceutical industry.

2.1   Methodology

In order to address these objectives, the chapter is structured along with two main issues:

1. Impacts in both local and global communities. These impacts relate to groundwater, surface water, aquatic ecosystems, drinking water, etc.
2. Solutions and measures intended to reduce discharges into waterbodies.

3   The City of Hyderabad

3.1   Hyderabad's Demography

The city of Hyderabad is currently the joint capital of the newly created state of Telangana and residual Andhra Pradesh. Hyderabad has an area of 217 sq. kms comprising of 16 administrative areas (called *mandals*), 1 town, and 67 villages with a population per the 211 census of 39.43 *lakhs* (hundred thousands) citizens. It is categorised as a district. The Musi River flows across the metropolitan city of Hyderabad, which is home to more than 10 million people.

As per the Government of Telangana data of 2016, around 30% of the state's urban population resides in the capital city of Hyderabad. 60 million people live within the Godavari river basin.

3.2 Hyderabad's Economy

Telangana contains several major manufacturing industries in bulk drugs, pharmaceuticals, agro-processing, cement- and mineral-based industries, high precision engineering, textiles, leather, apparels, automobiles and auto components, spices, horticulture, poultry farming, biotechnology, defence equipment and others. The Industrial Sector contributes at present around 25–30% to the Gross State Domestic Product and presents a steady growth dynamic.

As per the 2011 Census, the state's total population of the Telangana State is about 35 million. The majority of the state's population resides in rural areas and mainly depends on agriculture for their livelihood.

As per the GDDP estimates of 2015–16, the districts of Hyderabad, Rangareddy and Medchal-Malkajgiri occupy the top three positions with Rs. (rupees) 1363.88 million, Rs. 823.59 million and Rs. 476.04 million, respectively. These top three districts' GDDP (viz., Hyderabad, Rangareddy, Medchal-Malkajgiri), along with Sangareddy, constitute about 52% of Telangana's total GSDP, indicating that Hyderabad and its surrounding areas contains the state's economic hub.

The GSDP at current prices for the year 2016–17 (Advanced Estimates) is anticipated at Rs. 6542.94 million, exhibiting a growth rate of 13.7% over the previous year. Hyderabad's Gross District Domestic Product for the financial year 15–16 was Rs. 1360 million against the GSDP of Rs. 8430 million which amounts for 16% of the state GDP. The share of Telangana's economy in National GDP is 4.28% in 2016–17, as against 4.21% in 2015–16.

Hyderabad has the highest population density of 18,172 persons per sq. km as compared to the state average of 321 persons per sq. km. The literacy rate in Hyderabad is 83.25% as compared to a state average of 66.54%.

As per Annual Survey of Industries 2015–16, the numbers regarding factories registered under Factories Act operating in the top four districts are shown in Table 2.

420 of the 449 registered pharma manufacturers are concentrated in the three districts of Hyderabad, Rangareddy and Medak, which constitute almost 93% of

Table 2 Industrial factories working under the Factories Act and persons employed, 2015–16

District	Manufacture of chemical and chemical products	Manufacture of pharmaceuticals, medicinal chemicals and botanicals
Hyderabad	43	54
Medak	192	105
Rangareddy	92	261
Total of above three districts	327	420
Total of all districts in Telangana	395	449

Source Annual Survey of Industries 2015–16

total production—thus characterising the area as a pharma manufacturing hub (The Statistical Year Book 2017).

The three districts of Hyderabad, Rangareddy and Medak are home to the industrial areas of Jeedimetla, Pashamylaram (meaning pharma cluster), Bollaram and Patancheru. These industrial areas have discharge outflows into the Godavari River through the Manjira tributary.

The Manjira River flows approximately 15 kms away from the pharmaceutical hub of Bollaram in the Patancheru industrial area. The Nakavaggu, which is a tributary of the Manjira, is located hardly 5 kms away from this industrial area (Times of India, 29th Nov 2003, Industrial profile of districts).

Pharmaceuticals in Telangana constituted the state's highest export-related income in 2015–16, valued at 1283.7 million and amounting to 36% of total exports. Hyderabad alone contributes over 30% of India's pharmaceutical production (Reinventing Telangana, Socio-Economic Outlook 2017).

Telangana is among the leading states in pharmaceutical manufacturing and exports, second only to Maharashtra. The state accounts for around 25% of India's pharmaceutical manufacturing and exports. 43.4% of the state's manufacturing-related GDP in 2011–12 comes from the pharma sector. The sector contributes 4.79% of the state's GDP in 2011–12 at factor cost at market prices. Initially developed around Hyderabad in the Ranga Reddy District, industrial clusters rapidly developed in the Medak, Mahaboob Nagar and Nalgonda industrial areas. Some of most well-known international companies such as Dr. Reddy's Laboratories, Neuland Laboratories, Aurobindo Pharma, Hetero Drugs, Divis Laboratories, Mylan (now Matrix), Granules India, Virchow Laboratories, Glochem Industries Ltd., NATCO, Indian Immunologicals Ltd., Suven Life sciences, SMS Pharma, etc., have manufacturing facilities in the state (Govt of Telangana 2013–2014, Draft Export Strategy Framework of Telangana).

Hyderabad is well connected to all major cities of India by both air, rail and road. The Rajiv Gandhi International Airport connects Hyderabad to the major international cities. Hyderabad's economy is driven by pharmaceuticals, biotechnology and IT companies—including global giants. Hyderabad ranks first in the bulk manufacture of pharmaceuticals—it accounts for 40% of the total bulk production in India. Out of this 40%, about half is exported. Telangana contributes to over 35% of India's total pharmaceutical manufacturing. Hyderabad, the 'Bulk Drug Capital of India', accounts for almost 20% of Indian pharma exports, while the Telangana State contributes one-third of the country's total pharma production. There are around 400+ pharma companies in Telangana—including 170+ bulk drug units such as antibiotics—the majority of which operate in and around Hyderabad. Hyderabad as a pharma hub grew with the setting up of the Indian Drug and Pharmaceuticals Limited (IDPL) in 1961. The Jeedimetla Industrial Estate was created in 1977 and houses mainly the IDPL's ancillary units. The Bulk Drugs Manufacturers Association (BDMA) was formed in 1991 with headquarters in Hyderabad. The BDMA represents all bulk drug manufacturers in India and acts as a cooperative platform between the government and the pharma industry. The pharmaceutical companies located in

Hyderabad manufacture bulk drugs, fermentation products, synthetic drugs, inter-mediates, vitamins, vaccines, drug formulations, nutraceuticals, herbal products, speciality chemicals and cosmetics.

3.3 Local Impacts

Pharmaceutical companies located on the banks of the Godavari in the Adil-abad, Karimnagar, Warangal and Khammam districts all release untreated industrial effluents into this river (The Hans India, Oct 31, 2016).

Kazipally, Sultanpur and 15 surrounding villages on the banks of the Nakav-aggu rivulet—which joins the Manjira tributary and flows into the Godavari—are all affected as a result of effluent discharge by pharmaceutical companies (Vichi and Sahu 2016, Patancheru Industrial Area, https://www.ejatlas.org/print/patancheru-industrial-area-ap-india).

The residential areas of Vinayaknagar, Ayodhyanagar, and Birappanagar in the Quthbullapur constituency (up to Balanagar) have developed around the Jeedimetla Industrial Development area. A ten-km canal was built between Jeedimetla and Chinthal (covering the entire Quthbullapur constituency) that carries highly toxic wastewater from a group of chemical industries. The rain causes polluted water to flow from the unfenced canal into the residential area affecting the lives of the residents (Deccan Chronicle, 28th May 2017).

3.4 Around the Godavari River

The Godavari River is also known as the Ganges of South India and is the coun-try's second largest river. It rises in the Nasik district of the state of Maharashtra near Trimbak and flows eastward to join the Bay of Bengal in the state of Andhra Pradesh. The river is approximately 1465 kms long and has a total catchment area of 31 MHA (million hectares). The Manjira is one of the main tributaries of the Godavari River. A tiny rivulet called Nakavaggu joins the Manjira. This rivulet is heavily polluted by industries in the twin cities of Secunderabad and Hyderabad and supports no aquatic life at all. This polluted water is transported to the Godavari through the Manjira tributary. There are several industries in the Patancheru indus-trial area, mainly pharmaceuticals, which discharge their chemical wastewaters into this Nakavaggu rivulet.

3.5   Institutions Focusing on Environmental Regulation

The pharmaceutical industry has been classified in the Red Category by the Central Pollution Control Board (CPCB) due to the toxic and hazardous nature of its effluents. State Pollution Control Boards are tasked with preventing pollution as per the guidelines provided by the CPCB.

The Andhra Pradesh Pollution Control Board (APPCB) with its Head Office at Vijaywada consists of three zonal offices and nine regional offices covering the states of Telangana and Andhra Pradesh. The APPCB is the only Pollution Control Board in India that has introduced an Industrial Waste Exchange Programme. This is a free-listing service offered to industries based on the concept that waste generated by one industry can be of use to another. It also supports the concept of circular economy.

The APPCB has a list of various categories of both hazardous and non-hazardous waste that qualifies for co-processing along with a list of co-processing plants (cement manufacturers) which accept the classified waste. The waste generators list the sector they belong to along with the quantum of waste generated in Tons/annum (TPA). Details of the characterised waste—showing both the product manufactured and the waste generated—are available to the public. This serves to better facilitate the waste exchange process. Co-processing is a process through which both hazardous and non-hazardous waste is effectively disposed of through cement kilns. These operate at 1400 °C and use a lot of heat energy. The APPCB is also the only state that has a district-wise inventory of hazardous waste generation and issues permissions for co-processing hazardous waste.

With financial assistance from the World Bank, the Government of India—through the Ministry of Environment, Forest and Climate Change (MoEF&CC)—has started the implementation of a project called 'Capacity Building for Industrial Pollution Management Project' (CBIPMP). The objective is to establish a framework for 'scaling up the clean-up and rehabilitation of protected sites and facilitate the reduction of environmental and health risks associated with polluted sites'. This has been established as a National Programme for the Rehabilitation of Polluted Sites (NPRPS).

Some training programmes, mainly for the APPCB and CPCB, were held both in India and in the USA. The topics involved innovative remediation technologies and methodologies that focused on internationally recognised best practices, implementation of remediation processes based on cost-benefit analysis, use of modern tools intended to identify polluted sites and innovative remediation for contaminated sites. Two states, Andhra Pradesh and West Bengal, have been identified for undertaking remediation of hazardous waste in contaminated sites and municipal landfills as demonstration projects suitable for scaling up. The Andhra Pradesh Pollution Control Board is responsible for the project's implementation in the states of Andhra Pradesh and Telangana.

The APPCB under the World Bank's assisted project on the CBIPMP has set up an Environmental Compliance and Assistance Centre (ECAC) so as to achieve the objective of strengthening environmental compliance. The ECAC is a facilitation agency intended to assist industries establishing, operating and attaining regulatory and voluntary environmental compliance in a cost-effective manner by providing appropriate technological and legal/administrative information. It is an independent facilitator and sustainable organisation that provides consultancy and assistance specifically to micro-, small and medium enterprises as well as urban local bodies so as to improve compliance and competitiveness (Andhra Pradesh Pollution Control Board).

4 The Pharmaceutical Industry in Hyderabad

4.1 Types of Companies

Ever since the pharmaceutical industry in Hyderabad was established in the 1970s environmental pollution has affected agriculture and aquaculture in the region. Various newspaper reports highlight the conditions among local communities, from the reduction in milk output from farm animals to the difference in the visual appearance of local waterbodies. The various Industrial Areas in Telangana range from 15 to 2500 acres. The Telangana State Industrial Infrastructure Corporation (TSIIC) has 150 industrial parks which include sector-specific industries such as pharmaceuticals, biotechnology, IT, aerospace, apparels, automotive parks and special economic zones (SEZs). The Telangana State set up common effluent treatment plants (CETPs) at Jeedimetla and Patancheru and also a Treatment Storage Disposal Facility (TSDF) at Dundigal in the Jeedimetla Zone.

The Telangana State is the leading player in the pharma sector. Hyderabad, which is Telangana's capital, is well known as India's bulk drugs and vaccine hub. 33% of the bulk drugs exports from India come from Telangana. Hyderabad alone has over 400 manufacturers, spread over 12 industrial estates, with a limited scope to expand. The city has 245 USFDA-approved plants, the highest number in the world apart from the USA. Hyderabad has about 500 pharma and biotech companies.

With no further scope for expansion in Hyderabad, the state government is in the process of setting up Hyderabad Pharma City. Hyderabad Pharma City is a pharmaceuticals industrial park established in the Mucherla in Rangareddy district near Hyderabad. When completed, it will be the largest in Asia; its development follows global standards. The park is being set up on 19,330 acres in the Rangareddy district at a cost of Rs. 167.84 million crore. It is expected to attract Rs. 640.00 million investment and direct employment up to 170,000 people. The Expert Appraisal Committee (EAC) has highlighted that all bulk drug units in the proposed pharma city should have their own effluent treatment plants due to the high volume of effluent generation—with Zero Liquid Discharge (ZLD) facilities and including the requirement to reuse recovered water. The earlier proposed common effluent treatment plant (CETP) has not been accepted because the CETPs in the current industrial zones have proved ineffective and discharged hazardous waste—a tragedy which led to the shift of some manufacturing units from Hyderabad to 19 surrounding villages (The Hindu Business line, 5th October 2017).

4.2 Emissions Produced

On the other hand, treatment capacity of domestic sewages in India is far below the amount of sewage generated from 1.3 billion people. Only 31% of the total sewage produced (~38,254 million litres per day) in 908 cities were treated in 2008 (Subedi et al. 2015).

Pharmaceuticals are one of the main categories of emerging contaminants—besides steroids, oestrogens and personal care products. These contaminants, their degraded residues and transformation products are all cause for much concern (Ebele et al. 2017).

Ground water from an open well in southern India close to a water treatment plant receiving effluents from pharmaceutical production revealed the presence of ciprofloxacin, cetirizine, citalopram, enoxacin and terbinafine (Balakrishna et al. 2017).

The Patancheru Enviro Tech Limited (PETL) Waste Treatment Plant near Hyderabad, which received 1.5 MLD effluents from ~90 bulk drug manufacturers in the vicinity in Patancheru, was found to have the highest levels of pharmaceuticals ever reported in wastewater from anywhere in the world (Larsson et al. 2007).

With new analytical technologies such as gas chromatography (GC) and liquid chromatography (LC) used in combination with mass spectrometric detection, the gap in information regarding the presence, occurrence and environmental impact of these products is narrowing (Nikolaou and Lofrano 2012).

The transformation products of these pharmaceutical contaminants are not currently regulated. However, with new analytical methods providing information on their sources, occurrence and cycles—as well as risk assessment and eco-toxicological data—these contaminants will be subject to routine monitoring.

Gothwal and Shashidhar (2016) conducted a study intended to highlight the fact that improper effluent disposal of bulk drug manufacturing units can cause high levels of antibiotic contamination in a river. It was shown that the river in question could become an active site for the spread of AMR. Antimicrobial resistance (AMR) occurs when micro-organisms at the origin of infections not only survive exposure to a drug designed to eliminate them or prevent their growth but also evolve so as to become immune to said drug. Antimicrobial-resistant micro-organisms can travel around the world on a human host or through traded goods.

One of the major findings of Gothwal and Shashidhar (2016) shows that bulk drug industries monitor only conventional pollution parameters such as chemical oxygen demand/biochemical oxygen demand (COD/BOD) before disposing their effluent. There is no check on the amount of drugs in it; domestic sewage treatment plants and effluent treatment plants are not equipped to remove antibiotics. A focus on regulatory specification and safe disposal practices of bulk drug manufacturing units into surface water is a matter of urgency.

5 The Impact of the Pharmaceutical Industry's Emissions in the Godavari River

Pharmaceutical effluents discharged into the waterbodies leading to the Godavari River have far-reaching impacts on those in the vicinity of industrial zones where pharmaceutical plants are located: on the lives and livelihoods of local communities and on the surrounding ecosystem (specifically the aquatic environment), leading to the contamination of drinking water, health problems, reduction in milk output from livestock, etc. (Patneedi and Prasadu 2015).

The biological action by pharmaceutical compounds in humans themselves is better known than the toxic effects of hazardous chemicals discharged from manufacturing plants and industrial areas because pharma products were designed for human use. Characterisation of pharmaceutical compounds is important to assess potential harm. Since there is very little knowledge regarding the biological action or toxicity by pharmaceutical compounds on organisms, they were not designed for (non-target organisms), potential risks to non-target organisms exposed to pharmaceutical compounds in the environment is not yet entirely understood (Pfluger and Dietrich 2001).

Based on a study using an inter-laboratory mixture, the authors showed that it is possible to understand the combined effects of chemicals from water exposed to multiple mixtures. However, continuous development of bioanalytical and biomonitoring tools is needed so as to understand combination effects in vivo (Altenburger et al. 2018).

Cryptorchidism, which is the absence in the scrotum of one or both testes (usually because of the failure of the testis to descend) and hypospadias, which is the abnormal positioning of the opening of the urethra, are common birth defects of the male genitalia. These defects are the most frequently observed congenital malformations in boys caused due to androgen receptor (AR) antagonists. Humans are exposed to many anti-androgenic chemicals and earlier studies looked at effects of each chemical individually. These have shown no effect at very low concentrations. However, the study by Orton et al. 2014 has shown that the anti-androgenic effects of a mixture of AR antagonists—which come from a wide range of sources and exposure routes—have combined effects due to an additive effect present even at very low concentrations. It is concluded that the often low levels measured for individual AR antagonists are not a reliable indicator for dismissing risks from this class of chemicals (Orton et al. 2014).

Multi-component mixtures of steroidal pharmaceuticals, a phenomenon yet to be properly defined, are present in the aquatic environment. Fish are extremely sensitive to some of these steroids and sample pairs of fish were tested through exposure to five different synthetic steroidal pharmaceuticals. The data based on concentration-response on egg production showed a reduction in egg production of fish exposed to the mixture. The authors concluded that on its own each individual steroidal pharmaceutical in minute concentrations did not produce a statistically significant effect—yet the combined mixture did become what they termed as 'something from nothing' (Thrupp et al. 2018).

5.1 Impact on Drinking Water Sources

Surface waters have multiple stakeholders: communities and industries as consumers, governments and regulatory bodies as managers. The quality of surface waters has to be protected according to the needs and obligations of these multiple stakeholders (Houtman 2010; Smith 2014).

In Hyderabad, two among the six wells sampled close to the Nakkavagu and Isakavagu waterbodies—and which receive effluents from the Patancheru wastewater treatment plants—have been abandoned as a source of drinking water due to high presence of ciprofloxacin and cetirizine. However, the four remaining wells, although presenting these contaminants up to 1100 ng/L, are still being used so as to cater to local requirements (Fick et al. 2009).

Pomati et al. (2006) demonstrated through an in vitro research study that mixtures of pharmaceuticals—which included the most commonly found ciprofloxacin along with ibuprofen, ofloxacin, ranitidine and others at environmental concentrations—can inhibit the growth of human embryonic cells (Pomati et al. 2006; Rowney et al. 2009).

Even at very low concentrations, some pharmaceuticals have the potential to suppress an immune system's normal functioning (Kummerer and Al-Ahmad 2010).

Current research in this area reveals much cause for concern regarding potential risks to human health.

5.2 Impact on Aquatic Ecosystems

The issue attracting most attention was the presence of pharmaceuticals in river water, which was linked to the feminisation of fish found in downstream waters of wastewater treatment plants (Larsson et al. 1999).

This can affect the fish population and consequently the entire aquatic ecosystem—besides affecting the livelihoods and occupations of local fishermen.

Carlsson et al. (2009) investigated the effluent waters of the Patancheru wastewater plant which receives process waters from 90 manufacturers of bulk drugs in Hyderabad. It was found that even at effluent concentrations of 0.2%, there is a strong adverse effect on the growth of tadpoles—specifically through movement reduction and behaviour alterations. However, exposure to specific neurologically active drugs could also be the cause of this phenomenon.

5.3 Impact on Ground Waters

Two units manufacturing drug intermediaries were found to be illegally channelising effluents through underground pipelines into nearby drains. Water from these drains reaches Hussain Sagar and further ends in the Musi River. Locals have also informed Pollution authorities that groundwater was contaminated; in fact, this is a common problem in the Jeedimetla Industrial Area (Nilesh 2018, The New Indian Express, 6th April 2018).

Pharmaceutical manufacturing facilities contaminate nearby groundwater through the expel of various toxins. These toxins include heavy metals such as cadmium, vanadium, lead and arsenic, all of which were found in concentration levels much higher than the prescribed permissible limits for drinking water quality laid down by the World Health Organization (WHO) and the Bureau of Indian Standards (BIS) (Purushotham et al. 2017).

5.4 Health Impacts Among Local Communities

People in villages located on the banks of the Godavari River are often susceptible to illnesses due to the consumption of contaminated water. Fish, prawns and other aquatic species are threatened of extinction due to chemical contamination of river waters (Nordea 2016).

Pharmaceutical wastewaters have primarily been researched as regards contamination by individual constituents. More research is required as regards mixtures, wherein interactions can go up to quaternary levels. Further, studies on the effects of low doses for long periods of time are also needed (Synder et al. 2009, 2010).

Pharmaceutical waste products expelled into the aquatic environment become a blend of various heavy metals and organic pathogens that cause living organisms to become susceptible to potential mutagenic and genotoxic effects. Some studies have revealed that people living in the vicinity of pharmaceutical manufacturing industries suffer from water-borne diseases caused by water contamination. Some of these diseases include feto-maternal death, diabetes mellitus, gastroenteritis, impaired neurobehavioral effects, cardiac problems and hypertension (Ranjbongshi et al. 2016; Sharif et al. 2015).

Antibiotic contamination is widespread in waterbodies around manufacturing sites—and these issues have been neglected in India. Companies that manufacture bulk drugs are made to observe only the norms of Pollution Control Authorities, which only monitor conventional pollution parameters like chemical oxygen demand/biochemical oxygen demand (COD/BOD) analysed prior to effluent disposal. Domestic wastewater treatment plants and effluent treatment plants are not equipped with technologies necessary to remove antibiotics. Local communities are developing antibiotic resistance due to the proximity to waterbodies as well as consumption of water contaminated by these antibiotics (Bruni 2016).

5.5 Global Health Impacts

While communities living near waterbodies and production sites are paying an unreasonably high price within the global processes intended to bring affordable medicines to the world at large, health among global citizens is also being compromised due to the export of AMR.

Samples collected by researchers in the Changing Markets in 2016 were analysed at Oregon, USA for heavy metals and solvents commonly used in pharma production through the US EPA Method 8260 (a gas chromatographic method). This analysis revealed that substances relevant to pharmaceutical manufacturing are present at levels above maximum contaminant and safe exposure limits defined by the US' EPA, the Indian Bureau of Standards, the WHO, the European Union and the Californian EPA.

Samples were also collected in Sep 2017 from factories at the locations of Bachupally, Borpatla, Pashamylaram, Jeedimetla, Bollaram and Gaddapotharam and from waterbodies at Kazipally, Hussain Sagar Lake, Musi River at Adilabad and in the government-authorised hazardous waste management plant at Gaddapotharam. This analysis revealed that there were extremely high concentrations of heavy metals and solvents used in pharma manufacturing and organic synthesis. Most importantly, chromium VI (hexavalent chromium)—which is a human carcinogen due to its non-biodegradability and bio-cumulative capacity—was found in very high amounts in several samples. Natural attenuation of carcinogenic Cr VI into less harmful chromium substances is a natural phenomenon. However, the prevalence of high values of Cr VI indicates that untreated discharges containing Cr VI are an ongoing, continuous process over a long period of time (Nordea 2016).

The effluents from the Patancheru Effluent Treatment Plant (PETL) are discharged into the Isakavagu-Nakkavagu streams, which eventually flow into the Godavari River. Pharmaceuticals in the river and lakes are absorbed onto the soil/sediments, dilute, and undergo biological and/or photochemical transformations (Onesios et al. 2009). Pharmaceutical concentrations when measured 30 km downstream of the PETL showed significant reductions when compared to those observed at the PETL outlet; for example, cetirizine showed a reduction by 22 times as compared to the original value, while ciprofloxacin showed a reduction by 1400 times (Fick et al. 2009).

Shadreck and Mugadza (2013) highlighted the two sides of chromium (Cr). Cr (III) or trivalent chromium is an essential micronutrient required for carbohydrate, lipid and protein metabolism. By contrast Cr (VI), chromium's hexavalent form is a dangerous health hazard even in small doses. Exposure to high levels of chromium (VI) predominantly causes mutagenicity, carcinogenicity and teratogenicity.

The presence of hexavalent chromium in the natural environment is usually rare. It is associated with anthropogenic activities and industrial sources. Regular exposure to the hexavalent form of chromium causes tuberculosis, gastrointestinal bleeding, asthma, birth defects, stillbirths, infertility and skin ulcerations (Das and Mishra 2008; Palmer and Puls 1994).

Antimicrobial resistance or AMR is a broad term encompassing resistance to drugs that treat infections caused by other microbes as well, such as parasites (e.g. malaria or helminths), viruses (e.g. HIV) and fungi (e.g. Candida) (World Health Organization). AMR is the ability by micro-organisms to grow despite exposure to antimicrobial substances designed to inhibit their growth. These micro-organisms include bacteria, fungi, or protozoans. AMR is caused by the excessive use of antimicrobials, exposure to untreated or improperly treated drinking water and exposure to environments wherein resistant microbes are present. While any use of antimicrobials contributes to the development of AMR, unnecessary and excessive use of exacerbates it. The mixing of strains—which commonly occurs in wastewaters—allows said strains to share genetic material and thus transform into new, resistant strains at a faster pace. Technology and pharma developments need to overtake this pace so as to prevent the spread of AMR. In a review on AMR chaired by O'Neill (2014), AMR deaths by 2050 were estimated to reach 10 million—and the economic costs were estimated to reach 100 trillion USD, clearly affecting the labour force through mortality and morbidity. *Escherichia coli* (*E. coli*) and *Staphylococcus aureus* bacteria would increase the risk of tuberculosis and malaria, respectively. Loss of global GDP was estimated at 100.2 trillion USD; if left unchecked, AMR was estimated to reduce the world's output by 2–3.5% (O'Neill 2014).

The Patancheru-Bollaram cluster—where many drug manufacturers have had bases for over the last two decades—is a huge contributor to antibiotic resistance in bacteria as per the report by Nordea, 2016. According to local physicians, the breeding of resistant bacteria will affect the whole world.

In January 2017, the state government identified 1545 polluting industries to shift out of Hyderabad and earmarked to be relocated into 19 villages in the Medak, Nalgonda, Rangareddy and Mahbubnagar districts. The majority of the industries to be shifted are from pharmaceutical industry. Companies having Zero Liquid Discharge facilities have been allowed to stay. Central Effluent Treatment Plants will be set up in these villages so as to prevent shifting of pollution (The New Indian Express, 17th Jan, 2017).

Two reports (Nordea 2016 and Nordea 2017) show that although knowledge is available and awareness is present—and best available technologies notwithstanding—the key to deal with pharma pollution is to be found in the implementation of regulation using the polluter pays principle through an institutional setup combined with policy measures to control the present manufacturing discharges. This will close the loop and ensure a clean environment for the communities located around waterbodies.

It costs approximately Rs. 7000 to treat a tanker load of 10 kl with relatively low dissolved salts at the Patancheru Enviro Tech Limited (PETL), which is a CETP. Pharmaceutical manufacturing generates large volumes of effluents; costs will add up and discharging into drains will become a possibility.

Treated effluents from the PETL are also mixed with sewage; however, the cost of AMR treatment is exorbitant. In 2014, globally, 700,000 people die due to infections that were resistant to the usually prescribed antibiotics. By 2050, this number is estimated to reach 10 million. New research intended to discover more effective antibiotics adds to the costs of generated by an AMR epidemic (The Hindu, 18th Nov 2017).

It would seem that there could be an economic incentive to recover the large amounts of pharmaceuticals per litre of water being discharged. However, Larsson 2008) shows that in the case of ciprofloxacin, for example, the cost of purchasing bulk pharmaceuticals is often only a small fraction (1.5% of sales price) of the final pharmaceutical products in Sweden. A comparison between final prices and the investment and operating costs for producing a clean effluent may reveal the reason for the state of the water around pharma manufacturing sites.

6 Solutions for Reducing Emissions from the Pharmaceutical Industry

The detection of pharmaceuticals in water systems is the first step towards an understanding of what technologies can be used so as eliminate them. Advances in analytical technology are an important factor in the detection of pharmaceuticals.

6.1 Technology Options Available so as to Reduce Effluent Discharges

Technology options can turn a problem into an opportunity. This opportunity potential can lead to an exploring of new options so as to find solutions through a combination of technological means and actions by stakeholders.

The presence of Active Pharma Ingredients (APIs) in surface waters has been attributed to the discharge of wastewaters from pharmaceutical manufacturing—besides other sources such as domestic sewage and agricultural run-off.

API in the waterbodies in Hyderabad has been found to have reached alarming levels (Larsson et al. 2007).

Since pharma wastewaters are the major contributors of API products into surface waters, it is important to discuss the technology options available to this industry.

The technology applied during wastewater treatment needs to isolate and eliminate pharmaceuticals from the environment and not simply remove them through conversion—for example through the concentration of liquids into solid waste.

Technological Options Currently Used

The Common Effluent Treatment Plant (CETP)

The first common effluent treatment plant (CETP) in India was set up at Jeedimetla in Hyderabad in 1989 so as to address the effluent treatment of small-scale industries.

Common effluent treatment plants have been set up so as to treat wastewater from manufacturing plants that do not have their own dedicated effluent treatment plant or other treatment technologies due to financial reasons, lack of knowledge and skill or lack of space. One of Hyderabad's CETPs, the Jeedimetla Effluent Treatment Limited, contains facilities used to treat low TDS in combined wastewater treatment plant (CWWTP) and high TDS effluents in multiple-effect evaporator (MEE).

Low TDS effluents are subjected to a primary treatment and then processed into the biological treatment plant along with domestic wastewater within the activated sludge treatment process. Surface discharge standards are met and part of the treated wastewater is used for recycling/reuse; the remaining is disposed into sewers.

High TDS effluents are processed in multiple-effect evaporators after pre-treatment; the generated concentrate is processed in a spray dryer. The salt generated is then sent to an TSDF for further treatment and disposal. The condensate generated is processed in the biological treatment process.

Effluent Treatment Plants (ETP)

The most common and economical method for wastewater treatment used in pharmaceutical manufacturing plants is the biological treatment method. However, potentially hazardous constituents of wastewater—which may contain catalysts, organic solvents, recalcitrant compounds, reactants, intermediates and raw materials—make biological treatment difficult, resulting in lower removal efficiencies of chemical oxygen demand (Chelliapan et al. 2006). Enick and More (2007) estimated that up to 50% of wastewater from pharma plants is released without treatment. Biological processes are insufficient to remove hazardous constituents as these destroy the bacteria used and render the process ineffective.

Zero Liquid Discharges (ZLD)

The purpose of a well-designed Zero Liquid Discharge (ZLD) system is to reduce the volume of liquid waste requiring treatment and to produce a clean source of treated, recyclable water. A common ZLD approach is to concentrate wastewater through evaporation and crystallise the brine in high concentrations into a solid. Besides meeting the norms set by the Pollution Control Board, treating wastewater helps in recycling water within the manufacturing unit without discharging a single drop, hence the name Zero Liquid Discharge.

Zero Liquid Discharge has been mandated by the state government and some of the larger manufacturers have technology that does not dispose of any waste in liquid form as it is all converted into solids. This solid waste is incinerated or sent to government-authorised landfills.

In order to attain ZLD status, wastewater is first stripped off the low boiler solvents by a solvent stripping column and is either recovered for recycle or disposal. This process reduces the chemical oxygen demand of the wastewaters.

Reverse osmosis (RO) membrane systems are used for recycling water and the RO reject is sent to multiple-effect evaporators (MEE).

TDS streams are processed in MEE and the concentrate generated is spray dried. The salt generated is a mixed salt which is sent to a Treatment and Solid Disposal Facility (TSDF).

The economics of a Zero Liquid Discharge plant will depend on the optimisation of its operating costs, improvement of steam economy and increase in solvent recovery (Popuri and Guttikonda 2016).

ZLD has its own disadvantages in that the generation of solids can pose a disposal problem or a storage challenge. Also, the high cost of energy is a trade-off against the benefits that the ZLD approach provides. Stricter regulations for wastewater disposal, costly penalties for non-compliance, increasing wastewater costs sometimes outweigh the high expenses related to ZLD installation and operation. Countries such as the USA, China and India represent the major ZLD markets for dual reasons of large populations and need to conserve local water resources and ecosystems (Rajamani 2016).

In Hyderabad, only 86 out of the 220 bulk drug manufacturers have Zero Liquid Discharge facilities. (The Hindu dated 18th November 2017). A telephonic survey was conducted across pharmaceutical manufacturers in various industrial areas of Hyderabad so as to find out the technology options being used in order to manage wastewaters from pharmaceutical production. Out of 122 manufacturers contacted, 114 responded. The results are presented in Table 3.

The top five wastewater management technologies used by the pharmaceutical industry in Hyderabad are CETP (18% or 21 companies), ZLD (using RO + MEE + ATFD + landfill) with 18% of companies. In-house ETP was reported by 16 respondents (14% of the sample) followed by solvent recovery and recycling systems claimed to be used by 13 respondents (11% of the sample). Some of the raw materials are expensive and it is cost-effective to recover them. Similarly, valuable catalysts

Table 3 Technology options used for wastewater management across Hyderabad's pharma industry: a sample survey

Technology options used for wastewater management	No. of respondents	%
Central Effluent Treatment Plant (CETP)	21	18
Zero Liquid Discharge (using MEE)	21	18
In-house Effluent Treatment Plant (ETP)	16	14
Solvent recovery and recycling System	13	11
Product Recovery System	10	9
Source reduction (effluent reduction through process optimisation)	9	8
Green chemistry (Green Reagents, Biocatalysts, Green Solvents)	7	6
Membrane technology (reverse osmosis)	6	5
Solvent stripping system for disposal	4	4
Flow chemistry	2	2
Incinerator and Landfill	2	2
Continuous manufacturing	1	1
Soil Biotechnology (SBT)	1	1
Wet Air Oxidation Process	1	1
	114	100

are also being recovered by Product Recovery Systems. Product Recovery Systems have been adopted by 10 companies—making up 9% of the total respondents.

Some emerging technologies are green-chemistry based, Soil Biotechnology (SBT), and flow chemistry. These have yet to be adopted on a larger scale. Such technologies prevent the generation of waste in the first place (green chemistry-based) or reduce wastewater generation (flow chemistry or source reduction technologies).

Some Emerging Technologies

Soil Biotechnology (SBT)

This is a geo-filter system whose technology is based on a bio-conversion process wherein nature's fundamental reactions—namely respiration, photosynthesis and mineral weathering—are used so as to bring about the desired water purification. This is a biological process that supplies oxygen and therefore can treat industrial wastewaters of organic nature. SBT can be extended for attaining Zero Liquid Discharge for products expelled by the pharmaceutical industry. In the locations of pharmaceutical manufacturing where SBT is being used, water recovery is approximately 98% and no sludge is generated.

The initial costs are high; the construction of a plant with a capacity of purifying 10 lakh litres of wastewaters a day will cost Rs. 1.2 crore to Rs. 150 million. However, maintenance and operating costs are low—approximately Rs. 4 to purify 1000 litres (Times of India, 24th Jan, 2012).

Chemical oxygen demand (COD) and biochemical oxygen demand (BOD) removal of over 90% of expelled products can be achieved using this technology. And again, SBT has low annual operation and maintenance costs—comparable to land-based systems and lower than both conventional and most advanced technologies (Kamble et al. 2017).

Flow Chemistry

Flow chemistry provides a novel approach to conduct chemical synthesis in a continuous flowing stream instead of through traditional batch stationary reactors.

As the industry is facing increasing pressure on both quality and costs, the concept of continuous manufacturing becomes a pivotal technology intended to achieve more efficient, reliable and economic pharmaceutical production.

Pharmaceutical manufacturing is a process involving multi-step synthesis currently achieved through batches. Each step generates waste and impurities. Flow chemistry makes the flow step synthesis more efficient by generating fewer impurities which in turn reduces and simplifies downstream steps such as purification.

Flow chemistry in the pharmaceutical industry becomes a challenge because unit doses are very small and there is a stringent consistency of quality requirements. Continuous flow processes are currently being used only where hazardous and costly materials are present.

In flow chemistry, scale-up from microgram to kilogram quantities often requires minimal chemical modifications or reactor engineering.

While there are technological challenges for continuous processes in the areas of separation, purification, and integration of biocatalysis, companies working with flow chemistry expect to achieve a number of goals: reduction in cycle time; significant reductions in waste, carbon emissions, and raw materials; tighter impurity control; higher consistency in downstream processes; and greater overall yield (Thayer 2014).

Collaboration between pharmaceutical companies and equipment manufacturers is necessary to develop customised micro-reactors that meet the requirements for production scales (Cao 2017).

In a review paper by Bauman and Baxendale (2015), it was demonstrated that flow chemistry is an innovative technology that has enabled the efficient multi-step synthesis of 27 APIs. Current estimates suggest that there will be a 5–30% increase in flow chemistry in pharmaceutical manufacturing over the next few years—with industry and academia collaborating so as to achieve a higher degree of customisation as well as provide training for flow-based synthesis to the student body. Flow chemistry enables increased productivity, much lower capital employment and minimum waste generation (Baumann and Baxendale 2015).

Technologies allowing for a reduction at source

Source reduction is defined as any practice that:

– Reduces the amount of any hazardous substance, pollutant, or contaminant entering any waste stream or otherwise released into the environment (including fugitive emissions) prior to recycling, treatment or disposal.
– Reduces the hazards to both public health and the environment associated with the release of such substances, pollutants, or contaminants (US EPA 2015).

Deegan et al. (2011) explain in a review paper that although the majority of studies describe the elimination of parent pharmaceuticals as a removal, the disappearance of a compound need not necessarily mean that it has been completely removed. Biotransformation and mineralisation are ill-understood possibilities. The degree of biotransformation can thus far be known only as based on the information about metabolites or end products of mineralisation.

Finally, as newer compounds are being manufactured with little knowledge regarding their eco-toxicity, the problem of pharmaceuticals in wastewaters cannot be solved—merely by adopting end-of-pipe measures. Reduction at source measures such as replacement of critical chemicals by greener chemicals and reduction in raw material consumption need to be pursued as top priorities (Deegan et al. 2011).

Source reduction technologies are those that prevent waste generation at the source itself. It is a concept that provides an alternative to the traditional process of first generating waste and then trying to manage said waste through end-of-the-pipe technology solutions. Such alternatives would help reduce environmental impacts and above all present possibilities for resource recovery.

Recovery technologies

An equalisation tank as seen in effluent treatment plants is a holding tank used for the collection of wastewaters from various processes. The process of separation of compounds—if at all attempted—becomes much more technically difficult due to the mixing and interactions between various chemical compounds. The process also adds to the costs of treatment. Segregation of individual waste streams is the first step towards any recovery process. The characterisation of the various chemical components in any individual stream and not just lumped parameters such as pH, TDS and COD is the next step (Oliveira Júnior et al. 2013).

Once the stream's components are characterised, the process of identifying relevant technologies or technology platforms becomes easier.

Recovery technologies ensure environmental and economical sustainability for industrial wastewater management by recovering salts and organics and recycling the water back into the manufacturing process. Membranes of various types can be used at the source of the pollutants' generation and relevant as well as useful material can be recovered. Various other treatment technologies can be used so as to treat the concentrates generated from the process. Some technology platforms include crystallisation for recovery of salts; extraction for recovery of liquids (mainly organics); and dewatering for non-evaporative concentration of dilute streams.

Green chemistry approaches

Pharmaceutical companies are under an increasing pressure to reduce their environmental impact—specifically with regard to the copious amounts of waste and tainted water being generated from pharmaceutical production. Green chemistry is defined as 'the design of chemical products and processes that reduce or eliminate the generation of hazardous substances' (U.S. EPA 2015).

Green chemistry approaches work on the principles of green chemistry, which primarily focuses on waste reduction through a metric called the E-factor—total of waste produced [kg] per units of mass of product [kg]; water is normally excluded from the calculation (Sheldon 2010).

However, process mass intensity (PMI) is a metric that has been adopted by the pharmaceutical industry. PMI is the Total mass of materials used to produce a specified mass of product, measured units of mass of input materials (e.g. kg) per units of mass of product (e.g. kg). The calculation of the PMI is performed starting with commonly available materials

$$E \text{ factor} = PMI-1$$

Of the two metrics, PMI seems to be a better option as it indicates the overall greenness of a process as it considers all materials used whereas the E-factor only indicates the waste produced and the impacts resulting from rendering it harmless (Jimenez-Gonzalez et al. 2011).

Biocatalysis or Enzymatic technology

Drug manufacturers today are facing increasing market pressure to reduce development timelines and costs. At the same time, there is a need to provide safer and more effective drugs. The pharma industry is therefore looking for sustainable, effective alternative technologies. Stoichiometric reagents used in organic synthesis generate large quantities of by-products and dealing with these adds to the volume of waste created by the manufacturing process. Catalysis involving the right catalyst improves cycle times and minimises waste streams. As a consequence, product value is enhanced and costs are reduced. Catalysis has been around for a few decades now in the form of chemocatalysis, which includes metal catalysts and organocatalysts.

Due to the limitation of low efficiency attributed to high catalyst loading and long reaction times, as well as by the difficult process of separation of the catalysts from products, many of the known transition-metal catalysts and organocatalysts do not meet the requirements for modern industrial manufacturing processes (Zhou 2016).

Transformations that were not possible through the use of either catalytic system alone are now possible through the combination of transition metal catalysis and organocatalysis. This combination is able to improve the reactivity, efficiency and stereo-control of existing chemical transformations, although a few challenges still need to be addressed (Du and Shao 2013).

Greener routes towards pharmaceutical manufacturing are sustainable insofar as they consume fewer resources and produce less waste. Replacement of more traditional reactions mediated by transition metal catalysts with those facilitated by enzymes, or biocatalysts, is an option which is receiving growing attention. An emerging area exists regarding the replacement of more traditional reactions carried out by transition metal catalysts with those mediated by enzymes or biocatalysts. Biocatalysis or enzymatic catalysis provides pathways to implement green chemistry. The biodegradable nature of enzyme catalysts offers high selectivity and functional group compatibilities. Biocatalysis can progress under mild ambient or near-ambient temperatures using less solvents and a lesser number of steps. Biocatalysis is considered to be greener than transition metal catalysis as it meets the green chemistry principles of safety, atom economy, waste prevention and energy efficiency. Pharma companies estimate that anywhere from 25 to 75% of their pipeline can benefit from biocatalytic methods as they have helped develop cost-efficient processes while reducing environmental impacts (Challener 2016).

An example of catalyst applications in the pharmaceutical industry is the new catalytic process developed by Pfizer for the manufacture of pregabalin—which was able to decrease the reaction inputs by a factor of 5, and the use of solvents by 90%. The Merck brand has a successful application of a biocatalytic process demonstrated in the manufacture of its type 2 diabetes drug Januvia (sitagliptin). The new biocatalytic route developed by Merck/Codexis improved productivity by 56 per cent. Overall waste generation was reduced by 19% through this process (Sharma 2015).

Monitoring sensors

Sensors intended to monitor river pollution are an emerging technological option in India.

The region around the Godavari River will benefit from such an innovative sensor network installation. It is designed to be a boat-based mobile sensing platform intended to gather data on both bacterial and chemical pollution. It will possess a cloud-based data collection and real-time mapping system.

The aim is to detect and anticipate pollutants entering the Godavari River ad gauge their impact on the local environment. The Institute of Molecular Engineering at the University of Chicago is working on this forecasting system designed for cost-effectiveness (The Hindu, 11th July 2017).

Management options intended to reduce effluent discharges

While technology options are an important part of the objective of reducing effluent discharges, management options also need to be explored and exercised (Pruden et al. 2013).

Knowledge sharing

The processes used to manufacture pharmaceuticals are fairly similar across the world. Manufacturers in the 'developed world' need to share knowledge and best practices intended to eliminate pharmaceutical effluents and achieve non-toxic levels in the water (Schaaf et al. 2016).

Transparency in the supply chain

Today modern manufacturing and distribution systems involve complex, global supply chains. It is vital that every supplier ensures that environmental impacts linked to API manufacturing discharges are in line with standardised regulatory requirements. Datasheets on environment, health and safety are made available to all members of the supply chain so as to enable the assessment and management of risks associated with discharge of APIs (Caldwell et al. 2015).

India processes APIs imported from China for further processing and exports finished products to the USA and EU. The information regarding the origin of APIs used is incomplete on product packaging and information leaflets for patients. Increased transparency in the supply chain regarding the origin and consumption of antibiotics would help in control of production, use and disposal into the environment (Larsson and Fick 2009).

Influence by major pharma buyers

Major buyers of pharmaceutical products such as the National Health Services (NHS) and others can through their purchasing power exert a strong influence on pharmaceutical manufacturers and encourage compliance with good environmental practices for disposal. The inclusion of environmental criteria into good manufacturing practices (GMP) needs to be pushed forward so as to ensure compliance by manufacturing companies and respect for environmental health (Marchang and Hurley 2016).

Dual-track approach

Regular wastewater treatment alone is not able to handle the destruction of pharmaceuticals as these are designed to be stable in nature. Advanced wastewater treatment technologies increase costs and are energy-intensive. Dual-track approach is a more sustainable option. End-of-the-pipe solutions (downstream measures) need to be supported by source-control approaches (upstream measures) so as to enable sustainable solutions.

For example, a combination of an oxidation technique (e.g. ozone treatment) and a supplementary adsorption technique (e.g. activated carbon filtration) can achieve an average removal rate of between 80 and 90% for the removal of pollutants for sensitive receiving waters. Cost- and energy-efficient as well as sustainable methods need to be developed so as to remove pollutants from effluent wastewaters (Schaaf et al. 2016).

7 Conclusions and Recommendations

There are various treatment methods being adopted by the pharmaceutical industry. The compounds in wastewaters determine the type of wastewater management technology adopted.

It is extremely difficult for a single technology to completely eliminate the presence of pharmaceuticals from wastewaters as there is no one general behaviour among the various compounds present. A combination of technologies, or hybrid technologies which combine both biological and chemical processes, is becoming the most practical end-of-pipe method for wastewater reduction and treatment.

'Reduction at source' measures such as the replacement of critical compounds and the reduction of valuable raw materials such as catalysts and solvents need to replace end-of-pipe measures. The adoption of green chemistry approaches prevents the generation of wastewaters. Recycling compounds at the unit operation stage can help recovery and reuse of expensive compounds and reduce wastewater production. The characteristics of the wastewater determine the type of treatment technologies or the combination to be used.

Due to the high cost of raw materials draining into the pharmaceutical industry's effluent waters, the focus of attention should be on recovery technologies intended to enable the recovery of these materials and improve profitability. However, preventing the generation of pollution in the first place through waste minimisation and green chemistry principles represents a step further and requires a change in mindset linked to the pressing need to reduce the E-factor or process mass intensity of every process, as a higher E-factor means more waste and, consequently, more pollution.

Once technologies have been identified, further filtering can be achieved by analysing costs and benefits—taking into consideration competitiveness needs and any regulatory impediments.

With technology options available and research continuing in this area of pollution remediation, the monitoring mechanisms put in place by institutions such as the Pollution Control Boards need to be supported by both resource and manpower allocation strategies.

Finally, detecting a specific pharmaceutical contaminant is not enough. The overall quality of water should actively take into consideration all elements directly affecting environmental sustainability and human health.

Management options coupled with technology options can pave the way to exploring and exploiting synergies among the best available technologies, existing policies and targeted outcomes.

References

Altenburger R, Scholze M, Busch W, Escher BI, Jakobs G, Krauss M, Krüger J, Neale PA, Ait-Aissa S, Almeida AC, Seiler TB, Brion F, Hilscherová K, Hollert H, Novák J, Schlichting R, Serra H, Shao Y, Tindall A, Tollefsen KE, Umbuzeiro G, Williams TD, Kortenkamp A (2018) Mixture effects in samples of multiple contaminants—an inter-laboratory study with manifold bioassays. Environ Int 114:95–106. https://doi.org/10.1016/j.envint.2018.02.013

Andhra Pradesh Pollution Control Board. Available at http://appcb.ap.nic.in/

Ankley GT, Brooks BW, Huggett DB, Sumpter JP (2007) Repeating history: pharmaceuticals in the environment. Environ Sci Technol 41(24):8211–8217. https://doi.org/10.1021/es072658j

Balakrishna K, Rath A, Praveenkumarreddy Y, Guruge KS, Subedi B (2017) A review of the occurrence of pharmaceuticals and personal care products in Indian water bodies. Ecotoxicol Environ Saf 137:113–120. https://doi.org/10.1016/j.ecoenv.2016.11.014

Baumann M, Baxendale IR (2015) The synthesis of active pharmaceutical ingredients (APIs) using continuous flow chemistry. Beilstein J Org Chem 11:1194–1219

Brief Industrial Profile of Hyderabad District. Government of India, Ministry of MSME. Available at: http://dcmsme.gov.in/dips/hyd%20profile.pdf

Brief Industrial Profile of Medak District. Government of India, Ministry of MSME. Available at: http://dcmsme.gov.in/dips/medak.pdf

Brief Industrial Profile of Ranga Reddy District. Government of India, Ministry of MSME. Available at: http://dcmsme.gov.in/dips/RR%20%20dist%20profile_ap.pdf

Bruni P (2016) Impacts of pharmaceutical pollution on communities and environment in India. Researched and prepared for Nordea Asset Management by Changing Markets and Ecostorm. Available at: https://www.nordea.com/Images/35-107206/impacts%201-20.pdf

Bureau O (2016) Godavari reeking of pollution. The Hans India. Oct 31, 2016. Available at: http://www.thehansindia.com/posts/index/Khammam-Tab/2016-10-31/Godavari-reeking-of-pollution/261805

Caldwell D, Mastrocco F, Hutchinson T, Lohne R, Heijerick D, Janssen C, Anderson P, Sumter J (2008) Derivation of an aquatic predicted no-effect concentration for the synthetic hormone, 17. Ethinyl Estradiol Environ Sci Technol 42(19):7046–7054

Caldwell DJ (2016) Sources of pharmaceutical residues in the environment and their control. In: Hester RE, Harrison RM (eds) Pharmaceuticals in the environment, Vol 41—Issues in environmental science and technology. Royal Society of Chemistry, Cambridge, UK

Caldwell DJ, Mertens B, Kappler K, Senac T, Journel R, Wilson P, Meyerhoff RD, Parke NJ, Mastrocco F, Mattson B, Murray-Smith R, Dolan DG, Straub JO, Wiedemann M, Hartmann A, Finan DS (2016) A risk-based approach to managing active pharmaceutical ingredients in manufacturing effluent. Environ Toxicol Chem 35(4):813–822. https://doi.org/10.1002/etc.3163

Cao C (2017) Flow chemistry: pathway for continuous API manufacturing. Speciality Chemicals Magazine. Available at https://www.pharmasalmanac.com/articles/flow-chemistry-pathway-for-continuous-api-manufacturing

Carlsson G, Orn S, Larsson DGJ (2009) Effluent from bulk drug production is toxic to aquatic vertebrates. Environ Toxicol Chem 28(12):2656–2662

Challener C (2016) Going green with biocatalysis—enzymatic catalysis offers pharma manufacturers a way to implement the principles of green chemistry. Pharm Technol 40(8):24–25

Changing Markets Foundation (2018) Hyderabad's pharmaceutical pollution crisis: heavy metal and solvent contamination at factories in a major Indian drug manufacturing hub. Nordea and the Changing Markets Foundation. Available at: https://changingmarkets.org/wp-content/uploads/2018/01/CM-HYDERABAD-s-PHARMACEUTICAL-POLLUTION-CRISIS-FINAL-WEB-SPREAD.pdf

Chelliapan S, Thomas W, Sallis PJ (2006) Performance of an up-flow anaerobic stage reactor (UASR) in the treatment of pharmaceutical wastewater containing macrolide antibiotics. Water Res 40(3):507–516

Chunduri M (2003) Manjira faces pollution threat. Times of India. 29th Nov 2003. Available at: https://timesofindia.indiatimes.com/city/hyderabad/Manjira-faces-pollution-threat/articleshow/329239.cms

Das AP, Mishra S (2008) Hexavalent chromium (VI): environment pollutant and health hazard. J Environ Res Develop 2(3):386–392

Daughton CG, Ternes TA (1999) Pharmaceuticals and Personal care products in the environment: agents of subtle change. Special Report. Environ Health Perspect 107(Suppl. 6):907–938

Deccan Chronicle (2017) Toxic materials pollute Hyderabad nalas. May 28, 2017. Available at: https://www.deccanchronicle.com/nation/current-affairs/280517/toxic-materials-pollute-hyderabad-nalas.html

Deegan AM, Shaik B, Nolan K, Urell K, Oelgemöller M, Tobin J, Morrissey A (2011) Treatment options for wastewater effluent from pharmaceutical companies. Int J Environ Sci Tech 8(3):649–666

Drug resistance through the back door: How the pharmaceutical industry is fuelling the rise of superbugs through pollution in its supply chains, Nordea Asset Management, March 2016, Impacts of Pharmaceutical Pollution on Communities and Environment in India (report researched and prepared by Changing Markets and Ecostorm) http://www.nordea.com/en/responsibility/responsibleinvestments/responsibleinvestments-news/2016/New%20report%20on%20pharma%20industry%20in%20India.html

Du Z, Shao Z (2013) Combining transition metal catalysis and organocatalysis—an update. Chem Soc Rev 42:1337–1378. https://doi.org/10.1039/C2CS35258C

Ebele AJ, Abdallah MA, Harrad S (2017) Pharmaceuticals and personal care products (PPCPs) in the freshwater aquatic environment. Emerg Contam 3(1):1–16

Enick OV, Moore MM (2007) Assessing the assessments: pharmaceuticals in the environment. Environ Impact Assess Rev 27(8):707–729. https://doi.org/10.1016/j.eiar.2007.01.001

Express News Service (2017) Hyd's polluting industries to be moved to 19 villages. The New Indian Express, 17th January 2017. Available at: http://www.newindianexpress.com/cities/hyderabad/2017/jan/17/hyds-polluting-industries-to-be-moved-to-19-villages-1560441.html

Fick J, Soderstrom H, Londberg RH, Phan C, Tysklind M, Larsson DG (2009) Contamination of surface, ground and drinking water from pharmaceutical production. Environ Toxicol Chem 28(12):2522–2527

Gadipelly C, Pérez-González A, Yadav GD, Ortiz I, Ibanez R, Rathod VK, Marathe KV (2014) Pharmaceutical industry wastewater: review of the technologies for water treatment and reuse. Ind Eng Chem Res 53(29):11571–11592. https://doi.org/10.1021/ie501210j

Gothwal R, Shashidhar (2016) Occurrence of high levels of fluoroquinolones in aquatic environment due to effluent discharges from bulk drug manufacturers. J Hazard Toxic Radioact Waste 21(3). https://doi.org/10.1061/(ASCE)HZ.2153-5515.0000346

Government of Telangana (2013–2014) Draft export strategy framework of Telangana. Commissionerate of Industries. Commerce & Export Promotion. Available at: http://www.industries.telangana.gov.in/Library/EXPORTSTRATEGY.pdf

Government of Telangana (2017) Reinventing Telangana: looking back looking ahead. Socio-Economic Outlook 2017. Planning Department. Available at: https://www.telangana.gov.in/PDFDocuments/Socio-Economic-Outlook-2017.pdf

Hernando MD, Mezcuaa M, Fernandez-Alba AR, Barcelo D (2006) Environmental risk assessment of pharmaceutical residues in wastewater effluents, surface waters and sediments. Talanta 69(2):334–342. https://doi.org/10.1016/j.talanta.2005.09.037

Houtman CJ (2010) Emerging contaminants in surface waters and their relevance for the production of drinking water in Europe. J Integr Environ Sci 7(4):271–295. https://doi.org/10.1080/1943815X.2010.511648

Jimenez-Gonzalez C, Ponder CS, Broxterman QB, Manley JB (2011) Using the right green yardstick: why process mass intensity is used in the pharmaceutical industry to drive more sustainable processes. Org Process Res Dev 15(4):912–917. https://doi.org/10.1021/op200097d

Kamble SJ, Chakravarthy Y, Singh A, Chubilleau C, Starkl M, Bawa I (2017) A soil biotechnology system for wastewater treatment: technical, hygiene, environmental LCA and economic aspects. Environ Sci Pollut Res 24(8):13315–13334. https://doi.org/10.1007/s11356-017-8819-6

Koshy J (2017) Sensor network to map and predict pollution, effluents in Godavari, The Hindu. 11th July, 2017. Available at: https://www.thehindu.com/news/national/the-project-started-eight-months-ago-and-has-so-far-identified-two-hotspots-of-pollution/article19259008.ece

Kummerer K, Al-Ahmad A (2010) Estimation of the cancer risk to humans resulting from the presence of cyclophosphamide and ifosfamide in surface water. Environ Sci Technol 17:486–496

Larsson DGJ (2008) Drug production facilities: an overlooked discharge source for pharmaceuticals to the environment. In: Kummerer K (ed) Pharmaceuticals in the environment. Sources, fate, effects and risks. Springer, New York, NY, USA

Larsson DGJ (2014) Pollution from drug manufacturing: review and perspectives. Philos Trans R Soc B Biol Sci 369(1656):1–7. https://doi.org/10.1098/rstb.2013.0571

Larsson DGJ, Adolfsson Erici M, Parkkonen J, Pettersson M, Berg AH, Olsson PE, Forlin L (1999) Ethinylestradiol—an undesired fish contraceptive? Aquat Toxicol 45(2–3): 91-97. ISSN 0166-445X, E-ISSN 1879-1514

Larsson DGJ, de Pedro C, Paxeus N (2007) Effluent from drug manufactures contains extremely high levels of pharmaceuticals. J Hazard Mater 148(3):751–755. https://doi.org/10.1016/j.jhazmat.2007.07.008

Larsson DGJ, Fick J (2009) Transparency throughout the production chain-a way to reduce pollution from the manufacturing of pharmaceuticals? Regul Toxicol Pharmacol 53(3):161–163

Marchang S, Hurley N (2016) Drug resistance through the back door: how the pharmaceutical industry is fuelling the rise of superbugs through pollution in its supply chains. European Public Health Alliance (EPHA), 20 pp. Available at: http://epha.org/wp-content/uploads/2016/08/DRUG-RESISTANCE-THROUGH-THE-BACK-DOOR_WEB.pdf

Nikolaou A, Lofrano G (2012) Detection of transformation products of emerging contaminants. In: Lofrano G (ed) Green technologies for waste water treatment. Springer, Netherlands, pp 19–29

Nilesh V (2018) Has Telangana State Pollution Control Board unearthed only the tip of Jeedimetla pollution iceberg? The New Indian Express, 06th April 2018. Available at: http://www.newindianexpress.com/states/telangana/2018/apr/06/has-telangana-state-pollution-control-board-unearthed-only-the-tip-of-jeedimetla-pollution-iceberg-1797731.html

O'Neill J (2014) Antimicrobial resistance: tackling a crisis for the health and wealth of nations. The Review on Antimicrobial Resistance. https://amr-review.org/sites/default/files/AMR%20Review%20Paper%20-%20Tackling%20a%20crisis%20for%20the%20health%20and%20wealth%20of%20nations_1.pdf

Oliveira Júnior HM; Sales P; Oliveira DB, Schimidt F, Santiago MF, Campos LC (2013) Characterization and genotoxicity evaluation of effluent from a pharmacy industry. Rev Ambiente Água Interdiscip J Appl Sci 8(2):34–45. http://dx.doi.org/10.4136/ambi-agua.1107

Onesios KM, Yu JT, Bouwer EJ (2009) Biodegradation and removal of pharmaceuticals and personal care products in treatment systems: a review. Biodegradation 20(4):441–66

Orton F, Ermler S, Kugathas S, Rosivatz E, Scholze M, Kortenkamp (2014) Mixture effects at very low doses with combinations of anti-androgenic pesticides, antioxidants, industrial pollutant and chemicals used in personal care products. Toxicol Appl Pharmacol 278:201–208

Palmer CD, Puls RW (1994) Natural attenuation of hexavalent chromium in groundwater and soils: United States Environmental Protection Agency. EPA Ground Water Issue. Available at: https://www.epa.gov/sites/production/files/2015-06/documents/natatt_hexavalent_chromium.pdf

Patneedi CB, Prasadu KD (2015) Impact of pharmaceutical wastes on human life and environment. Rasayan J Chem 8(1):67–70

Pfluger P, Dietrich DR (2001) Effects of pharmaceuticals in the environment—an overview and principal considerations. In: Pharmaceuticals in the environment. Springer, Berlin, pp 11–17

Pomati F, Castiglioni S, Zuccato E, Fanelli R, Vigetti D, Rossetti C, Calamari D (2006) Effects of a complex mixture of therapeutic drugs at environmental levels on human embryonic cells. Environ Sci Technol 40:2442–2447

Popuri AK, Guttikonda P (2016) Zero liquid discharge (ZLD) industrial wastewater treatment system. Int J ChemTech Res 9(11):80–86. ISSN(Online):2455-9555

Pruden A, Larsson DGJ, Amézquita A, Collignon P, Brandt KK, Graham DW, Lazorchak JM, Suzuki S, Silley P, Snape JR, Topp E, Zhang T, Zhu T-G (2013) management options for reducing the release of antibiotics and antibiotic resistance genes to the environment. Rev Environ Health Perspect 121(8):878–885. https://doi.org/10.1289/ehp.1206446

Pulla P (2017) The superbugs of Hyderabad. The Hindu. 18th Nov 2017. Available at: https://www.thehindu.com/opinion/op-ed/the-superbugs-of-hyderabad/article20536685.ece

Purushotham D, Linga D, Sagar N, Mishra S, Vinod GN, Venkatesham K, Saikrishna K (2017) Groundwater contamination in parts of Nalgonda District, Telangana, India as revealed by trace elemental studies. Geol Soc India 90:447–458

Rajamani S (2016) Novel industrial wastewater treatment integrated with recovery of water and salt under a zero liquid discharge concept. Rev Environ Health 31(1):63–66. https://doi.org/10.1515/reveh-2016-0006

Rajbongshi S, Shah YD, Sajib AU (2016) Pharmaceutical waste management: a review. Eur J Biomed Pharm Sci 3(12):192–206

Reddy AS (2017) The statistical year book 2017. Directorate of Economics and Statistics Government of Telangana Hyderabad. Available at: https://www.telangana.gov.in/PDFDocuments/Statistical-Year-Book-2017.pdf

Rowney NC, Johnson AC, Williams EJ (2009) Cytotoxic drugs in drinking water: a prediction and risk assessment exercise for the thames catchment in the United Kingdom. Environ Toxicol Chem 28(12):2733–2743

Sayadi MH, Trivedy RK, Pathak RK (2010) Pollution of pharmaceuticals in environment. J Ind Pollut Control 26(1):89–94

Subedi B, Balakrishna K, Sinha R, Yamashita N, Balasubramanian V, Kannan K, (2015) Mass loading and removal of pharmaceuticals and personal care products, including psychoactive and illicit drugs and artificial sweeteners, in five sewage treatment plants in India. J Environ Chem Eng 3:2882–2891

Schaaf N, Karlsson J, Borgendahl J, de Pedro C, Fiedler E, Flygar H, Göthberg P, Lonaeus K, Magnér J, Mattson B, Olsen T, Olsson B, Schultz S, Svedberg A, Svinhufvud K (2016) Water and pharmaceuticals—a shared responsibility. Working paper Nr. 26. SIWI, Stockholm

Shadreck M, Mugadza T (2013) Chromium, an essential nutrient and pollutant: a review. Afr J Pure Appl Chem 7(9):310–317

Shalini K, Anwer Z, Sharma PK, Garg VK, Kumar N (2010) A review on pharma pollution international. J PharmTech Res 2(4):2265–2270. ISSN: 0974-4304

Sharif A, Ashraf M, Anjum AA, Javeed A, Altaf I, Akhtar MF, Abbas M, Akhtar B, Saleem A (2015) Pharmaceutical wastewater being composite mixture of environmental pollutants may be associated with mutagenicity and genotoxicity. Environ Sci Pollut Res 23(3):2813–2820. https://doi.org/10.1007/s11356-015-5478-3

Sharma V (2015) Applicability of Green chemistry in Pharmaceutical processes. Pharma Bio World. pp 34–36. Available at: http://piramalpharmasolutions.com/wp-content/uploads/Applicability-of-Green-Chemistry-in-Pharmaceutical-Processes.pdf

Sheldon RA (2010) Introduction to green chemistry, organic synthesis and pharmaceuticals. In: Dunn PJ, Wells AS, Williams MT (eds) Green chemistry in the pharmaceutical industry. ISBN: 978-3-527-32418-7. https://doi.org/10.1002/9783527629688.ch1

Smith Jr JS (2014) Health impact of pharmaceuticals in the environment and in drinking water: potential for human and environmental risk. In: Goldstein WE (ed) Pharmaceutical accumulation in the environment: prevention, control, health effects, and economic impact. Taylor & Francis Group. ISBN: 978-1-4665-1745-5

Snyder S, Lue-Hing C, Cotruvo J, Drewes JE, Eaton A, Pleus RC, Schlenk WD (2009) Pharmaceuticals in the water environment. NACWA. Association of Metropolitan Water Agencies. Available at: https://www.acs.org/content/dam/acsorg/policy/acsonthehill/briefings/pharmaceuticalsinwater/nacwa-paper.pdf

Snyder SA, Trenholm RA, Bruce GM, Snyder EM, Pleus RC (2010) Toxicological relevance of EDCs and pharmaceuticals in drinking water. Environ Sci Technol 44(14):5619–5626. https://doi.org/10.1021/es1004895

Snyder SA, Trenholm RA, Bruce GM, Snyder EM, Pleus RC (2008) Toxicological Relevance of EDCs and Pharmaceuticals in Drinking Water, Awwa Research Foundation, Denver, CO

Thayer AN (2014) End to end chemistry. Chemical and Engineering News. 92(21):13–21. Available at: https://cen.acs.org/articles/92/i21/EndEnd-Chemistry.html

Thrupp TJ, Runnalls TJ, Scholze M, Kugathas S, Kortenkamp A, Sumpter JP (2018) The consequences of exposure to mixtures of chemical: something from 'nothing' and 'a lot from little' when fish are exposed to steroid hormones. Sci Total Environ 619–620:1482–1492. https://doi.org/10.1016/j.scitotenv.2017.11.081

Tischler L, Buzby M, Finan DS, Cunningham VL (2013) Landfill disposal of unused medicines reduces surface water releases. Integr Environ Assess Manag. 9:142–154

TNN (2012) IIT alumnus shows natural way to process waste water. Times of India. Jan 24, 2012. Available at: https://timesofindia.indiatimes.com/city/chennai/IIT-alumnus-shows-natural-way-to-process-waste-water/articleshow/11611412.cms

US EPA (2015) Basics of green chemistry, US Environmental Protection Agency, Washington, DC, USA. Available at: https://www.epa.gov/greenchemistry/basics-green-chemistry#definition

Vichi L, Sahu G (2016) Patancheru industrial area, meta information. Available at: https://ejatlas.org/conflict/patancheru-industrial-area-ap-india

Zhou Q (2016) Transition-metal catalysis and organocatalysis: where can progress be expected? Angew Chem Int Ed 55(18):5352–5353. https://doi.org/10.1002/anie.201509164

Chapter 4
Ground Water Irrigation in a Contested Space: A Tale of Technological Change, Institutional Transformation, and Co-option

Poulomi Banerjee and Leon M. Hermans

1 Introduction

The sustainability of local water management systems is shaped by the interaction of different components, including institutions, natural resource characteristics, technology options, and socio-economic aspects. At the local level, especially the way in which local communities use and manage water resources has been studied since the early 1990s, looking at the interplay among the local natural resource base, livelihood activities amid the citizenry and institutional regimes (Schlager and Ostrom 1992; Agrawal and Gibson 1999; Sikor et al. 2017). A bit more recently, also the importance of a new space within community resource management has emerged—along with the growth of urban areas and the growing importance of peri-urban spaces (Allen 2003; Kombe 2005; Narain et al. 2013; Narain and Singh 2017; Gomes and Hermans 2018). Peri-urban areas, typically located at the fringe of urban areas (and defined as the areas at the interface between expanding cities and rural areas), exist in a constant state of flux, characterized by transition more than by well-defined spatial parameters (Narain and Singh 2017). In many cases, peri-urban areas consist in previously rural areas now undergoing a rapid process of transformation into more urban-based communities, increasingly linked to the nearby city.

Whereas community resource management is by itself a complex phenomenon, the complexity further increases in a peri-urban area. Especially when it comes to the institutional dimension, the state of transition that characterizes peri-urban areas poses a challenge, as we observe a high level of heterogeneity among both actors and dynamics (Gomes et al. 2018).

P. Banerjee
SaciWATERs, Hyderabad, India
e-mail: poulomi@saciwaters.org

L. M. Hermans (✉)
Delft University of Technology & IHE Delft, Delft, The Netherlands
e-mail: l.m.hermans@tudelft.nl; l.hermans@un-ihe.org

© Springer Nature Switzerland AG 2020
S. Bandyopadhyay et al. (eds.), *Water Management in South Asia*,
Contemporary South Asian Studies, https://doi.org/10.1007/978-3-030-35237-0_4

Peri-urban areas are given increasingly scientific and policy attention and rightly so. Strategies and processes of socio-economic development present complex challenges and opportunities to all actors involved. The most negatively affected are probably peri-urban farmers as degrading surface water flows, industrial pollution, and declining groundwater tables all affect their choice of technologies, management practices and, above all, their decision to either cooperate with or contest current policies (Vij and Narain 2016). Collective action takes a new form—reflected in new networks and ties which determine farmers' access to one of the most critical local natural resources: groundwater.

In this chapter, we will look at the developments in a specific irrigation system, as well as at the networks and trust relations, control mechanisms, rights, forms of collective action, and social exclusions—and at how all these dynamics are mediated through power politics. The case is contextualized in a peri-urban space of a rapidly growing delta city—Kolkata. Many peri-urban areas witness ever-expanding urban processes which lead to shrinking rural landscapes and put much pressure on traditional agriculture-based livelihoods. One common trend is that irrigated agriculture is more and more restricted to high-value commercial crops, using groundwater as a source for irrigation. This phenomenon is also seen in the deltaic space around Kolkata, which is endowed with relatively rich groundwater reserves. Here, tube well irrigation and village-level farmer associations have played a critical role in harnessing groundwater resources and surviving in the face of urban development. Yet, the inter-linkages between local farmers' technological choices and actual agricultural purposes are often blurred. There are several angles from which farmers collectively come together, exercise mutual trust, build up new ties and networks, and modify existing ones so as to effectively share and manage this crucial resource.

We argue that understanding these trajectories requires a better understanding of local communities as heterogeneous social structures (Agrawal and Gibson 1999) and benefits from the use of an updated schema of property rights (Sikor et al. 2017).

2   Theory

Theoretical understandings hold peri-urban communities as diverse and heterogeneous groups, with different sectors and actors positioned differently on the power ladder (Allen 2003; Arabindoo 2009; Narain et al. 2013; Vij and Narain 2016; Narain and Singh 2017). The peri-urban is regarded as a political space wherein actors with diverse and often conflicting interests pursue a range of different goals. In such a space, collective action critically depends on the power play, negotiation strategies, and management approaches by different players. Each player and group of players need to be viewed in a more heterogeneous scale wherein behaviour is influenced by several socio-political forces operating both internally and externally (Gomes et al. 2018).

It is clear that understanding the complex interactions within peri-urban communities and the ways these use local resources against changing socio-political landscapes requires a break from earlier works on community-based use of natural resources. Said earlier works considered a given community as a homogeneous unit and its collective action and working towards progressive social change (Agrawal and Gibson 1999). We instead adhere to the reasoning presented by the seminar paper by Agrawal and Gibson (1999), which stresses the importance of recognizing local communities as heterogeneous groups which vary by size, composition, norms, and resource dependence, and whose decisions are mediated by different institutional arrangements. Hence, Agrawal and Gibson (1999) suggest that a focus on institutions was needed for a better understanding of community-based natural resource management.

Institutions are a critical element within the complexity of peri-urban communities. Institutions are generally understood to be the "prescriptions that humans use to organize all forms of repetitive and structured interactions" (Ostrom 2005: 3). These create relatively stable patterns of interaction, which help actors to establish expectations about each other's behaviour and about the collective norms and sanctions that should guide behaviour (Ostrom 2005).

Existing water management institutions were not developed to meet the needs of peri-urban communities. They were developed according to either rural or urban sets of rules—and specifically for the management of water systems on a physical basis (in the case of a River-Basin Authority or Irrigation Management Group). Peri-urban communities face new and different realities not contemplated in previous policies—for instance, as regards the use and value of land and water resources and as regards the stakeholders and social structures involved. Power balances may shift—for instance, when local real estate developers come into play or when local industries or other urban market forces affect local developments, but also when migration flows occur both into and out of the peri-urban communities. As a result, existing institutions may no longer be functional or credible–and communities begin searching for new ways to facilitate their collective action (Gomes and Hermans 2018). The result may take the form of negotiated action, power struggles, conflict, or cooperation (Narain and Singh 2017). Therefore, the institutional dimension requires specific attention in a peri-urban context, arguably even more so than in other settings.

Of particular importance to the use and management of groundwater resources in dynamic peri-urban communities is the notion of access and property rights. Who has the right to use the resource at stake, and to what purpose? How do current developments influence technological and livelihood choices among local users? For the past decades, property rights for natural resources have been studied using a framework proposed by Schlager and Ostrom (1992) in which five types of rights were identified: access, withdrawal, management, exclusion, and alienation (Ostrom 2010). Recently, Sikor et al. (2017) proposed a modification and refinement of this framework, adding another level of rights and additional types: transaction and monitoring rights; instead of access and withdrawal, they proposed rights to direct versus indirect benefits. We expect that this updated framework may be well-suited to the specific context of peri-urban areas, a premise we will further explore in this chapter.

3 Methodology

We use a single, in-depth case study in our research. This case was identified as part of a larger research project on groundwater management in peri-urban areas within the Ganges' delta. Research sites were identified based on a quantitative cluster analysis and a rapid rural appraisal in promising villages (Banerjee 2016). Further data for the selected research sites was then collected through a household survey (Banerjee and Jatav 2017).

In this chapter, we focus on one village near Kolkata, India, wherein we conducted an additional field campaign. Twenty-five farmers, including cultivators and leaseholders, participated in semi-structured interviews and discussions conducted between November 2017 and May 2018 in order to explore issues surrounding groundwater irrigation, access, and management. Additional interviews were conducted with water professionals and practitioners as well as members of local authorities in order to elicit environmental, political, and legal factors affecting land and water grabbing, exploitation, access, and management.

The interviews were conducted in the agricultural fields at an early hour of the day and typically lasted around 1.5–2 h. All participants were heads of household, male, and between the ages of 30 and 74. Conversations were typically conducted in Bangla and translated into English. Interviews sought to elicit responses in three categories:

(a) What urbanization trends you have seen operating in your village and how have they determined your choice of irrigation technologies for the past 15 years?
(b) To what extent are your inter-personnel relationships and your connections with both land and water determined by your access to resources?
(c) To what extent there is a transformation in the relations of power and trust regarding the access to land and water?

The qualitative data was analysed in accordance with the research questions at hand and organized according to each respondent's relationship with both land and groundwater—either as cultivators or as tenant farmers—from head and tail end of the irrigation network. The respondents practised diverse cropping patterns, from mono-cropping of paddy to the cultivation of fresh vegetables and horticulture. Farmers taking part in the main deep tube well groundwater irrigation system were divided into two main categories—cultivators and tenant farmers. From each of these groups, specific respondents were selected based on their location (head and tail farmers) and also according to crop types practised—such as paddy growers, mixed cultivation of both paddy and vegetables, and garden farmers. 5% of respondents fell outside the deep tube well's catchment area. All the different farmers groups were collectively analysed in order to elicit the effects on resource accessibility across farming groups.

Using the thus collected and analysed study material, we focus on three specific issues. First, we describe and analyse how urbanization processes and socio-political forces influence farmers' choices of irrigation technologies and management strategies in such a contested space. Second, we discuss how these processes and strategies create new forms of trust and networks among farmers. Third, we look into the ways in which such transformed trust networks and associations reinforce power relationships in the process of co-opting and sharing the most valued local resource—groundwater.

4   Results

4.1   Introduction to the Case Study Village

Our focus is on a small deep tube well farming community in one of Kolkata's peri-urban villages. We aim to analyse differential access to and use of groundwater resources. The case study village has been selected for its proximity to Kolkata city, for the presence of groundwater irrigation and for its status as a typical agrarian-based community undergoing rapid change in land use due to peri-urban development.

The village's farming community is homogeneous in terms of its socio-religious identity, yet political subjectivities and the connections it entertains with more powerful political actors present heterogeneity along its power axes. The village has an 80% Muslim population which has permanently settled since the Bengal's partition. This population is composed of traditional agriculturalists and holds a maximum share of agricultural holdings. Middle-class Hindus form 15% of the population and are mostly employed in the service sector. The remaining 5% of the populations are recent migrants from Bangladesh and several Indian states and engage as wage labourers in the manufacturing sector.

The village community's social composition can be seen in its settlement patterns, as each section has distinct religious identities, economic status, and livelihood patterns (See Fig. 1). Hindus reside in clusters 2 and 3 (Fig. 1) and are major commuters between the village and Kolkata city. They hold very small parcels of land in and around neighbouring villages (Table 1). Muslims occupy 1 and 4, while migrants reside in the most congested and polluted areas—cluster 2.

At the time of study, unrest among farmers over land acquisition and degradation, compensation, and industrial pollution resulted in protests and sentiments of uncertainty. Five participants interviewed staged protests at the panchayat office and in front of a local, dying factory. A case between farmers and residents vs said dying factory was logged in Kolkata's High Court. The state of concern about falling groundwater levels was reflected in the discussions with both villagers and professionals. Overexploitation and low recharge were mentioned several times during the interviews.

The studied village lies about 25 km^2 from Kolkata's municipality and is well connected with the Kalyani express highway and with the Barasat and Jessore roads. Feeder roads such as the Mahispatta and the Madhyamgram provide easy market linkages. The village is in its early stage of de-intensification, wherein agriculture is slowly losing its competitiveness in relation to other sectors and farmers are reacting by reducing their main agricultural inputs and livestock. Agriculture presents a mixed

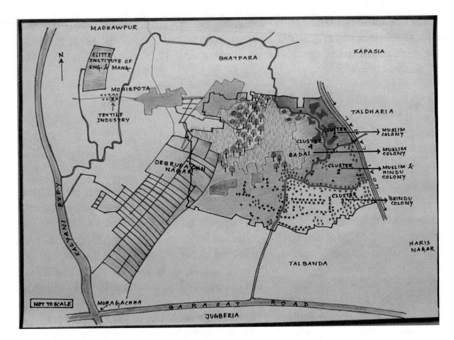

Fig. 1 Social map of case study village near Kolkata. *Courtesy* Prosanta Kumar modal and Samrat Biswas

Table 1 Livelihood distribution across religious groups

Percentage distribution of sample households by primary and secondary source of income						
Sector	Primary			Secondary		
	Hindu	Muslim	Total	Hindu	Muslim	Total
Growing of crops, market gardening, horticulture	4.3	45.7	38.3	0.0	13.3	10.9
Forestry, logging, and related services	0.0	1.0	0.8	0.0	1.0	0.8
Manufacturing and repairing	65.2	16.2	25.0	26.1	16.2	18.0
Construction	17.4	18.1	18.0	0.0	8.6	7.0
Transport, storage, and communication	8.7	8.6	8.6	8.7	3.8	4.7
Trade (wholesale and retail)	4.3	4.8	4.7	26.1	11.4	14.1
Education, health, and social work	0.0	1.9	1.6	4.3	0.0	0.8
Real estate, renting, and business activities	0.0	1.0	0.8	0.0	0.0	0.0
Other services	0.0	2.9	2.3	8.7	1.0	2.3
None	0.0	0.0	0.0	26.1	44.8	41.4
Total (%)	100.0	100.0	100.0	100.0	100.0	100.0
Total (numbers)	23	105	128	23	105	128

Source Compiled from household surveys, Banerjee and Jatav (2017)

space wherein small-size, high-value outputs such as vegetables and fruits share land with traditional paddy farms. There are evidences of non-farm diversification, mostly relating to construction and transports.

The village itself is a traditional paddy growing area and an important source of fresh vegetables and fruits for local markets as well as the large wholesale markets of both Panihati and Kolkata. The village's uniqueness lies in the presence of a deep tube well irrigation system that still manages to operate against the strong forces of urbanization. The 900 ft deep tube well, jointly managed by the irrigation department and the beneficiaries, narrates the story of rights, control, and power play between state and non-state actors. The nexus between industries, real estate groups, local authorities, and farmers has implications not only on the village's small and shrinking agricultural sector but also on the larger social consciousness, political subjectivity, norms, trust, and conceptions among each of these groups. This in turn determines the way resources are accessed and managed. Understanding the trajectory of irrigation in a peri-urban context thus requires interpreting these relationships and their connection with wider socio-political processes. The present case study aims to unravel the inter-linkages between politics and policies and the influence these have on farmers' technological choices and management strategies for the last 15 years.

4.2   How Urbanization and Socio-Political Forces Affect Farmers' Choices of Irrigation Technologies

Currently, irrigated agriculture in the village depends solely on the use of one deep tube well. This has not always been the case. During conversations about the history of irrigation in the village, cultivators and lease farmers spoke about different surface irrigation techniques. Lift irrigation through diesel motors from the community ponds and *Noai Khal* (natural surface water channel) flowing all along the village formed the major sources before the year 2000. About 5% of respondents spoke about private, shallow tube wells, either individually owned or community-managed. Water sharing was informal. Aman paddy was the principal crop, while cultivation of boro vegetables was limited to a few large farmers who owned private, shallow tube wells. The first shift from diesel-run, private shallow tube wells into a single electricity-driven, public deep tube well occurred with the advent of industrial forces.

Economic reforms by a leftist government created opportunities for micro-enterprises and self-employment, which were reflected in a massive spread of small-scale manufacturing industries in the village. The village now has 50 odd dyeing factories, mostly illegally constructed, reportedly extracting on average 10,000 to 15,000 gallons of water per hour. Most of these factories possess more than one bore well; each such well has a horsepower capacity of 7.5. Almost all respondents spoke about illegal encroachments and dumping of industrial waste on the Noai Khal and thus eventually to the agricultural fields. Loss of agricultural land due to inundation has been reported by 70% of respondents. The conjunctive use of surface water and groundwater was also limited. Such loss, coupled up with unlimited extraction of groundwater by industries, drastically reduced farm sizes, increased fragmentation, and lowered aquifer levels. Respondents reported falling groundwater levels for the last 15 years, most critically felt since 2013. The village has lost the shallow layer between 150 and 180 ft. Between 1998 and 2000, Aman rice production showed a mediocre performance and there was limited expansion of water control.

Such changes increased inequalities among farmers, wherein a few large owners benefitting from lump-sum compensation and political connections became richer while the vast majority of farmers experienced great disadvantage—70% of respondents narrated their present struggle for basic needs. Nevertheless, all the respondents reported poor access to irrigation facilities in the middle of the 1990s. In 2000, there was an attempt to revamp the village's irrigation status and make groundwater available to the broad masses of peasantry; to this end, a deep tube well was installed by the state. The catchment area was 300 bighas (a measure of land area varying locally from 1/3 to 1 acre or 1/8 to 2/5 hectare) with 150 beneficiary farmers. The then Panchayat President, along with a few powerful farmers, took the initiative to donate land for laying the network and construction of pump houses. The network has 12 chambers, interwoven by channels built with concrete and mud and reaching out to the tail farmers. The major shares of the chambers reportedly went to the Panchayat President and his brother. Table 2 shows the distribution of chambers across head farmers (using anonymized names).

Table 2 Distribution of chambers across head farmers

Chamber	Farmer[a]	Land holding size (bigha = a measure of land area varying locally from 1/3 to 1 acre or 1/8 to 2/5 hectare)
1	P. Amar	Ex panchayat Pradhan, 7 bighas
2	P. Amar	
3	M. Mohammad	12 bighas
4	A. Amar	Over 10 bighas
5	A. Amar	Brother of Ex Panchayat Pradhan. At present works as dealer, has sold some lands to real estate company, rest has been given as lease
6	Parvan	60 bighas
7	C. Amar	20 bighas
8	D. Amar	10 bighas
9	E. Amar	7 bighas
10	F. Amar	6 bighas
11	G. Amar	12 bighas
12	Waleed	10 bighas

Source Compiled from field 2018
([a]fictional names)

The deep tube well, installed by the government's Irrigation Department and jointly managed by the beneficiaries and said Department, soon became the lifeline of the village. Yet urban forces continued to shape the pathway of this irrigation system. 40% of the farmers interviewed reported their struggle against powerful industries, land sharks, and real estate giants. For example, farmers sold 30 bighas of land to a multi-millionaire real estate company, reducing the cultivable land to 150 bighas—while unused land increased from 20 to 100 bighas over the last 7–8 years.

Falling groundwater levels are exhibited in the way boring has been deepened for the public tube well as well as for private wells. Our discussion with the plumber involved in the construction of tube wells for households and industries revealed that water levels fluctuated between 50 and 80 ft in peak season. Falling depth of the aquifer can be seen in changes of the diameter of the casing and depth at which cylinders and screens are placed in tube wells. The diameter of the casing that houses inlet, cylinder, and piston valves has increased to 9 inches, while the depth at which cylinders are placed has increased to 60 ft. Submersible pumps are generally placed at about 80 ft be placed deeper—to 180 ft. The situation has become more alarming in the last 5 to 7 years.

The deep tube well itself was deepened several times. The first boring was done two months after its initiation. The then Finance Minister Sri Ashim Dasgupta took the initiative in deepening the bore and changing its motor. In 2004, a second deepening was done wherein the depth increased to 600 ft. Last year, in 2017, the motor was replaced and the depth of the boring was further increased to 900 ft. The duration and number of watering per crop have increased as it now takes longer to cover one cropped area.

The uncertainty associated with falling groundwater tables is further accentuated with the presence of water filtration plants in and around the village. Limited access to both land and water forced the farmers to look for alternate sources. Ten per cent of respondent farmers from the tail end of the irrigation system mentioned the presence of water extraction from storage tanks through siphon machine or lever pumps (of 40 horsepower) in order to irrigate fields. About five per cent of the farmers outside the deep tube well system also reported purchasing water for irrigation purposes from the system's head farmers.

At first glance, it is apparent that irrigation technology is subjected to the acquisition of resources by powerful urban players. However, it is the larger political agenda that determines whether there will be change in this course. Until around 2013, lift irrigation from natural streams and ponds was used conjunctively with the deep tube well by large tenant farmers as well as those of tail end and by outsiders. However, excessive pollution of surface channels and lack of government initiatives intended to dredge and clear such channels significantly limited their use. Since then, irrigation solely relies on the deep tube well system. Thus, the choice of technology is a forced decision among farmers with limited alternatives as crucial resources are controlled by the state and powerful urban players. It is the complex interplay between the physical complexity, plurality of water rights, and state control that alters the traditional notion of access and reconfigures the relations between resources and users.

4.3 How This Creates/Transforms Trust and Networks Among Farmers

Urban spread as well the control and extraction of both land and water resources has impacted not only technological choices but also farmers' perceptions of and responses to such forces. Urban control coupled up with political consciousness influenced the way farmers organize—individually and collectively—so as to harness resources. The formation of networks, lobbies, and groups has become critical in order to access resources. New individual connections and trust networks are created and often previously existing ones are reshaped—elucidating power dynamics, rights, and forms of control. In this context, the tube well system exemplifies the nexus between farmers, political parties, industries, and real estate giants. It depicts how individual networks and trust-based relations come into increasing tension among communities in the tube well command area. One interface where such complex interplay of power groups and formation of networks can be observed is the management of a beneficiary committee of tube well users.

West Bengal's Left Front's policy of making irrigation participatory—with a larger role played by communities—resulted in the 1980s in the formation of user groups. The state government had several misgivings regarding the efficiency of state-managed irrigation systems and decided to hand over the administration of tube wells to the Panchayat Samiti (local self-government). The Krishi Sangha Village is one such beneficiary committee, created as a watchdog for the public deep tube well installed in 2000. The committee was responsible for the operation, maintenance, and collection of water-related taxes. Constituent members were selected from the micro-irrigation department and the Gram Panchayat (local village council). Active involvement by the then Panchayat Pradhan (Council President) and the willingness by small and marginal farmers helped in commissioning the tube well. During the initial years, the committee was presided by the Panchayat Pradhan, which totalled 5–7 members. The tube well operator was selected from the Irrigation Department, and his role was to distribute water and maintain proper records regarding the tube well running, water rates charged, and revenues realized.

In 2011, a new political party came into power, and a new, informal government position regarding a *private* tube well operator was created from among the beneficiary group. The position, although at the time portrayed as "political" and meant to assist the public operator, became much more than it was intended to be. It became a channel through which political parties tried to control both access to and use of resources. It also became a key link in the interaction between urban players and farmers, who would negotiate and form lobbies to control both land and water. Since 2011 till date, this new position has been continuously occupied by a single person— one enjoying strong political connections with the ruling party. This person designs water distribution patterns and solves disputes regarding water sharing from the deep tube well. The position became highly contested and reveals considerable political interference. 50% of the respondents reported corruption, bribing, and unwarranted practices in the way this assistant waterman functions. Water distribution norms of

Table 3 Operation of the village deep tube well

Tube well operator	Year of operation	Important events since its commission	Details
Oper. 1	2000	The first boring was done two months after its initiation. The then Finance Minister Sri Ashim Dasgupta took the initiative to deepen the bore and change its motor	
	2003	Tube well failure	Depth increased to 600 ft
	2005		
Oper. 2	2006		
	2008	Tube well failure	New machine installed
	2009		
Oper. 3	2009		
	2014		
Oper. 4 and Oper. 5	2014		
	2016	Tube well failure	Further deepened to additional 35 ft
Oper. 3	2016		
	2017		Motor replaced and depth increased to 900 ft
Oper. 3	2017		
	2018		

Source Compiled from the field; operator names removed and replaced by numbers

"first come first served" are often violated, and any attempt to remove the operator's hegemony quickly becomes a violent fight (Table 3).

This not only increased the uncertainty in the amount and timing of water received by different farmers but also created factions among the farming community. Two groups of farmers, each with distinct political affiliations and associations with the operators, emerged. The axes of differentiation are further deepened with the interference by urban players, particularly factories and real estate giants. A small network of farmers and tube well operators supported by the state, industrial houses, and land sharks de facto control the major portion of both land and water.

4.4   How Transformed Networks Reinforce Power Relations

The restructuring of networks and power flows has two very clear-cut implications. One, it made large tenant cultivators vulnerable—their investments were high, and so was uncertainty regarding water supplies. The village has few large tenant cultivators with long-term lease agreements and who cultivate boro paddy, vegetables, and mangoes. Land acquisition and blockage of the surface channels limited alternate irrigational sources. Since around 2013, these farmers had no other option but to depend on the deep tube well. The situation became more critical for those who were unable to network with the private tube well operator or who had a political affiliation with opposition (as in not in power) political parties. Our discussion with tenant farmers revealed change in the power dynamics, with a greater concentration of power in the hand of a few small and medium cultivators associated with the operator.

The water supply became the operator's monopoly. In spite of having marginal holdings in the command area, the operator seems to manage to control a large group of farmers. This person not only decides on water distribution but also allegedly sells water illegally to the tail-end farmers and to those outside the tube well command area. Unlike the official, flat water tax regarding the deep tube well, rates in this illegal market vary. The water operator solely decides on water charges and often tail-end farmers are forced to pay very high sums. Group discussions with some tail-end farmers and those falling outside the ambit of the command area affirm and confirm the growing power of the operator.

Two, changing networks, associations, and lobbies resulted in the emergence of a new, third group of farmers—in addition to the established groups of tenant farmers and farmers who owned their own land. This new group consists of land brokers and is composed either by ex-farmers or by small and marginal cultivators. The farmers in this new group are well connected to the government on one hand and to industries on the other. They have a privileged position in the information network—often being invited to public agricultural extension meetings wherein new support schemes are announced and explained—and they are often involved in land transactions. The farmers in this small third group not only control information flows, but also reap benefits of their position at the expense of farmers. Not surprisingly, this is reported to undermine trust networks and undermine collective action. A small but significant part of respondents shared grievances regarding lack of information and poor representation in agrarian events such as Krishi Mela.

5   Discussion

The trajectory of groundwater irrigation in the case study village shows how the emergence of new local industries and real estate developments—linked to the growth of the urban centre of Kolkata—combined with the availability of groundwater irrigation technology, in particular the installation of a deep tube well, to determine the evolution of agriculture and water use. This trajectory illustrates the importance of new types of property rights, as proposed in the updated conceptual framework by Sikor et al. (2017).

5.1 Authoritative Rights

Authoritative rights of definition and allocation determine the control rights. The original scheme proposed by Schlager and Ostrom (1992) did not distinguish these particular types of rights. However, we see clear instances of those two types of rights in our case study, most visibly in relation to the operator of the groundwater well used for irrigation. Upon the installation of the deep tube well by the then State Government in 2000, a government-appointed public operator was to exercise control rights over the well's groundwater. As mentioned, in later years (2011), when a new party came to power, this right was given to a private operator.

5.2 Control Rights

The exercise of control rights of management, exclusion, transaction, and monitoring is critical in our case. Especially management rights, the right to regulate the use and transformation of the resource are of critical importance. Originally, when shallow groundwater resources could be accessed by multiple farmers, and when surface water flows could be used as a fallback option for irrigation, these rights were somewhat dispersed among community actors. In the past few years, however, the decline in groundwater tables and the pollution of surface water created an exclusive dependence on a single deep tube well, whose management rights are concentrated in the hands of a private operator. Informally and "de facto", this has also given this actor the right to define who is to benefit from use rights and under what conditions—namely for what prices in the illegal water market. This means that management, exclusion, and transaction rights have now in the hands of one and the same actor. Monitoring rights—i.e. the right to monitor the use and the state of the resource—seem ill-defined; visible tensions and protests are mounting, but no clear resolution seems attainable today.

Furthermore, transactions of land and of government-sponsored agricultural support schemes are also important in our case. They are, to the dismay of some farmers, concentrated in the hands of a newly emerged group of "transaction farmers". These are farmers who no longer rely on farming as their primary source of income but who benefit from their intermediary position between local farmers on the one side and government and industrial parties on the other.

5.3 Use rights

Use rights in our case clearly result from changing property rights in the framework's higher order. The use of direct and indirect benefits is critically important, but they are determined by the higher-order rights.

An important issue related to these use rights relates to the right to dump industrial waste and polluted wastewater into surface water streams. Although we have not included this as a main point of attention in our analysis, industrial pollution clearly has played a key role in the village's irrigation trajectory. Since 2013, surface water sources are too polluted to be used in agriculture, dramatically augmenting local dependence on a single groundwater source. It appears that local industries have appropriated use rights—namely the right to use surface water streams to transport their waste—because of an inadequate handling of control rights. Either there were no formal rights defined for industrial waste, or these rights were not enforced.

6 Conclusion

Our case study shows that local groundwater management institutions are visibly changing in the peri-urban village under study. This was not so much a result of a consciously orchestrated government policy as a result of a process wherein government interventions—such as the installation of a private tube well or the determination of control rights in the hands of a public or private operator—interact with developments in the private market and the informal/illegal economy. All these forces shape local groundwater management.

Understanding such developments requires a property rights framework that goes beyond a government–community dichotomy and instead focuses on local communities as heterogeneous social groups. Furthermore, it requires an appropriate property rights framework that has sufficient detail and fits the peri-urban context. The property rights framework used in this study, proposed by Sikor et al. (2017) as a significant update to an earlier framework by Schlager and Ostrom (1992), provides a useful format to capture the dynamics in groundwater use, institutions, and technologies in the peri-urban environment around Kolkata.

Acknowledgements The research reported here is part of the NWO UDW Shifting Grounds project, funded by the Netherlands Organisation for Scientific Research under the grant W.07.69.104. The authors are grateful to Manoj Jatav, Prosanta Kumar Modal, Samrat Biswas, and Priya Sangameswaran for their contributions, support, and advice, as well as our colleagues from The Researcher in Kolkata for support in fieldwork and data collection.

References

Agrawal A, Gibson CC (1999) Enchantment and disenchantment: the role of community in natural resource conservation. World Dev 27(4):629–649
Allen A (2003) Environmental planning and management of the peri-urban interface: perspectives on an emerging field. Environ Urban 15(1):135–148
Arabindoo P (2009) Falling apart at the margins? neighbourhood transformations in peri-urban Chennai. Dev Chang 40(5):879–901

Banerjee P (2016) Report on case study selection. WP/SG/01/2016. Shifting grounds working paper series. SaciWATERs, Hyderabad

Banerjee P, Jatav M (2017) Thematic paper on urbanisation and groundwater use. Socio-economic system mapping. WP/SG/05/2017. Shifting grounds working paper series. SaciWATERs, Hyderabad

Gomes SL, Hermans LM (2018) Institutional function and urbanisation in Bangladesh: how peri-urban communities respond to changing environments. Land Use Policy. https://doi.org/10.1016/j.landusepol.2017.09.041 (In press)

Gomes SL, Hermans LM, Thissen WA (2018) Extending community operational research to address institutional aspects of societal problems: experiences from peri-urban Bangladesh. Eur J Oper Res 268(3):904–917

Kombe WJ (2005) Land use dynamics in peri-urban areas and their implications on the urban growth and form: the case of Dar es Salaam, Tanzania. Habitat Int 29(1):113–135

Narain V, Singh AK (2017) Flowing against the current: the socio-technical mediation of water (in) security in periurban Gurgaon, India. Geoforum 81:66–75

Narain V, Khan MSA, Sada R, Singh S, Prakash A (2013) Urbanisation, peri-urban water (in)security and human well-being: a perspective from four South Asian cities. Water Int 38(7):930–940

Ostrom E (2005) Understanding institutional diversity. Princeton University Press

Ostrom E (2010) Beyond markets and states: polycentric governance of complex economic systems. Am Econ Rev 100(3):641–672

Schlager E, Ostrom E (1992) Property-rights regimes and natural resources: a conceptual analysis. Land Econ. 249–262

Sikor T, He J, Lestrelin G (2017) Property rights regimes and natural resources: a conceptual analysis revisited. World Dev 93:337–349

Vij S, Narain V (2016) Land, water & power: the demise of common property resources in periurban Gurgaon, India. Land Use Policy 50:59–66

Chapter 5
Flood-Prone Ghatal Region, India: A Study on Post-'Phailin' Inundations of 2013

Nabendu Sekhar Kar and Sayantan Das

1 Introduction

A flood is an overflow of river or lake water that submerges land (CRED 2013). Floods occur mostly as a result of heavy rainfall events—whenever watercourses do not have the capacity to carry the excess water involved. However, floods can result from other phenomena—for example a coastal inundation resulting from a tropical cyclone, from a tsunami, or from a spring tide leading to above-average gauge levels (GA 2011). River floodplains and low-lying coastal areas are the most susceptible to flooding. Areas experiencing unusually long periods of heavy rainfall are also vulnerable (NDA 2013). The consequences of a flood may vary greatly—depending on its location and extent and on the vulnerability and value of both natural and man-made environments affected. Asia's tropical regions have experienced the highest number of floods due to high-magnitude tropical cyclones and monsoon downpours. These areas have accounted for more than 80% of all major floods occurring between 1900 and 2012 (Sanyal et al. 2013). Floods—including extreme events such as flash floods and torrential rains—have become increasingly common in India in the last few decades (Goswami et al. 2006).

Almost 42% of West Bengal's total area is susceptible to floods (Bandyopadhyay et al. 2014). Between 1960 and 2000, there occurred 13 high-magnitude floods (inundating over 11% of the state's territory) and 10 medium-magnitude floods (WBDMD 2014). Flood characteristics are different in different physiographic zones within West Bengal. The low-lying areas of the West Medinipur, East Medinipur, Howrah and Hooghly Districts—all of which are part of the Chhota Nagpur Plateau-fringe

N. S. Kar (✉)
Department of Geography, Shahid Matangini Hazra Government College for Women, Purba Medinipur, India
e-mail: nabendu@matanginicollege.ac.in

S. Das
Department of Geography, Dum Dum Motijheel College, Kolkata, India
e-mail: sayantdas@gmail.com

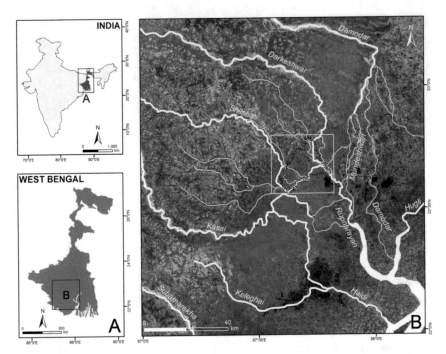

Fig. 5.1 Maps of the Ghatal region under different perspectives along with a general overview of the Bhagirathi–Hooghly River's western tributaries. *Notes* The area demarcated by black dots within the yellow rectangle in map **B** identifies the Ghatal CD Block. *Image* **B** GeoEye-1, 18.04.2016

fan system—often become flooded due to its basin-shaped topography (Bandyopadhyay et al. 2014; Figs. 5.1 and 5.3). Embankment breaches are common and result in a considerable delay in the drainage and splaying of suspended particles across the floodplains (Chapman and Rudra 2007; Bandyopadhyay et al. 2014; DoIW–GoWB 2014). In October 2013, the cyclone 'Phailin' befell the Chhota Nagpur Plateau region of Jharkhand and the western part of West Bengal, leading to severe floods.

2 Study Area

The Ghatal region is located in the West Medinipur District of West Bengal. Ghatal town (22.663° N 87.736° E) is the headquarters of the Ghatal Subdivision, which is one of the West Medinipur District's four subdivisions. The Ghatal Community Development (CD) Block, the concerned study area, has the largest area (216 km^2) amongst the five CD Blocks under the subdivision (Figs. 5.1 and 5.2). It is bounded by the Hooghly District in the north and east, the Dasapur-I and Dasapur-II Blocks of the West Medinipur District in the south, and the Chandrakona Block of the West Medinipur District in the west. At 1016 persons per km^2, the population density of

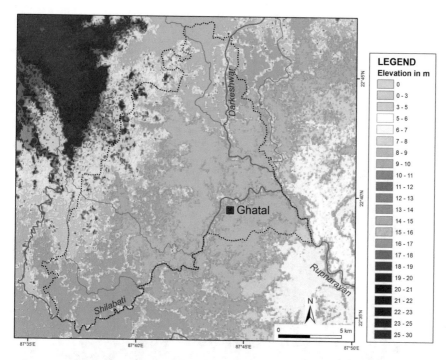

Fig. 5.2 A digital elevation model (DEM) showing the physiographic setting of the Ghatal CD Block and surroundings (boundary demarcated by black dots). *Note* The location of Ghatal town in the eastern (right) bank of the Shilabati River is also shown. The elevation of the plateau-fringe fan system is 10–30 m, whereas the elevation range near the confluence of the rivers is 6–9 m. The general slope of the region follows an axis from northwest to southeast. *Image* 1 arc-second (30 m) Shuttle Radar Topography Mission DEM (v-3: tile n22-e87), February 2000

Ghatal CD Block is slightly lower than that of West Bengal—which with an average of 1029 persons per km^2 is the second most densely populated state in India (CoI 2011). The population of Ghatal town was 54,591 in 2011 (CoI 2011). The Ghatal region receives plenty of rainfall every year between June and September (mean annual rainfall of 153 cm), when the southwest-led monsoon is most active (DoIW–GoWB 2014). During the post-monsoon months (October–November), tropical cyclones originating in the Bay of Bengal often bring torrential rains to this part of West Bengal. Traversed by a dense network of rain-fed, semi-controlled, east-flowing rivers such as the Shilabati, the Dwarkeshwar and the Rupnarayan (Figs. 5.1, 5.3 and 5.4), the Ghatal CD Block frequently becomes flooded.

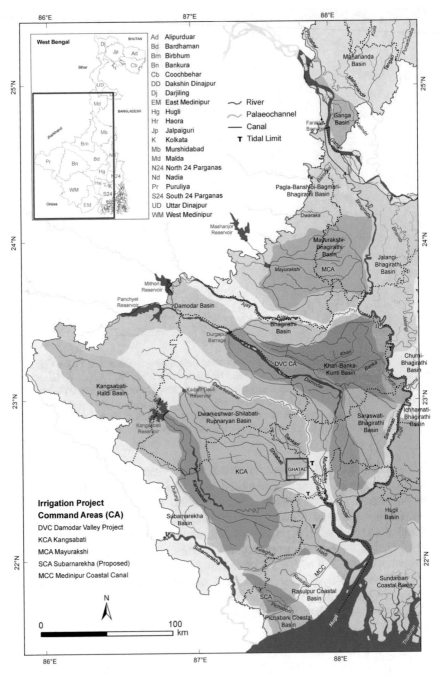

Fig. 5.3 Drainage map West Bengal's western areas (area marked in the index map) showing major river basins and the boundaries of installed irrigation projects. *Note* The red square on the map indicates the Ghatal region itself. After Bandyopadhyay et al. (2014)

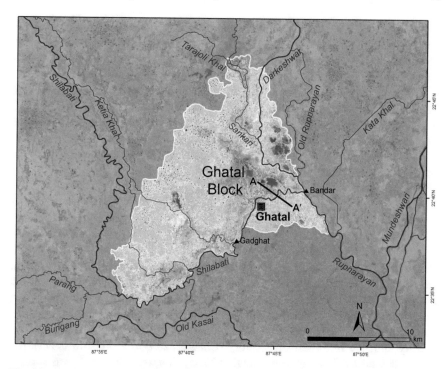

Fig. 5.4 Network of rivers around the Ghatal CD Block (enhanced subset). *Notes* All these rivers eventually drain into the Bhagirathi–Hooghly. The main direction of these flows runs towards the east and southeast. The location of two gauging stations (Gadghat for the Shilabati River and Bandar for the Rupnarayan River) is also shown. The extent A–A' indicates the profile shown in Fig. 5.7. *Image* Landsat 8 OLI (Path-139, Row-44), 16.11.2014

2.1 Geological and Physiographic Setups

The western part of West Bengal evolved as a coastal unit within the north-eastern Indian craton, which broke loose from the eastern Gondwanaland in the Early Cretaceous Era (133 Ma BP) and proceeded to drift northwards (Lawver et al. 1985; Powell et al. 1988; Duncan 1992). The east-flowing rivers from the Chhota Nagpur Plateau formed coalescing deltas on the continental shelf—the Bengal Basin—which continued to progress into the sea until the formation of the wider Ganga Delta during the Quaternary Period (Niyogi 1975; Agarwal and Mitra 1991). At present, these rivers—the Mayurakshi, Ajay, Damodar, Rupnarayan and Kangsabati–Haldi Rivers—are the major western tributaries of the Bhagirathi–Hooghly.

Alongside the degraded plateau, the landscape of West Bengal's western part is also characterized by the plateau-fringe palaeodeltas which resemble a subdued fan system (Bandyopadhyay et al. 2014). Based on geomorphic observations, this fan system merges with the westernmost and oldest part of the Ganga Delta in the east (Fergusson 1863; Bagchi 1944; Niyogi 1975), wherein the regional gradient lowers quite suddenly. Consequently, the major rivers within this region are separated into many branches, and their valleys have become narrower towards their respective downstream. The Ghatal region is located at the lower edge of the western palaeodelta complex wherein the regional gradient becomes much lower still (approximately 1 in 7989), than is the case in the upper part of palaeodelta (approximately 1 in 1003). The Ghatal town and surrounding areas' elevation range is between 7 and 9 m. It is here that the east-flowing Shilabati and Dwarkeshwar Rivers merge to form the Rupnarayan River (Fig. 5.2). The floods in these areas are caused by heavy downpours in the catchment areas of these rivers and subsequent water stagnation owing to the very low regional gradients present.

2.2 Drainage Characteristics

Originating in the Chhota Nagpur Plateau region, rivers such as the Mayurakshi, the Ajay, the Damodar, the Dwarkeshwar, the Shilabati, the Rupnarayan and the Kangsabati (Kasai) constitute western tributaries of the Bhagirathi–Hooghly River. These rain-fed rivers are known for creating havoc during the monsoon months. The Damodar—the largest of these rivers in terms of discharge—used to flood frequently until the mid-twentieth century (Chandra 2003). In order to prevent the misery caused by flooding, the Damodar Valley Corporation (DVC) was formed in 1948—in accordance with a suggestion by a search committee formed by the erstwhile British Government (DFEC 1943). In the south of the DVC's command areas during the 1950s, the Kangsabati Project (KP) was put in place by the West Bengal's Department of Irrigation and Waterways (DoIW–GoWB) so as to regulate water flows in the West Medinipur, East Medinipur, Bankura, Purulia and Hooghly Districts. As a part of these two projects, a number of reservoirs were constructed in upstream areas of both the Damodar and Kangsabati (Kasai) rivers and their tributaries (Bandyopadhyay et al. 2014; Fig. 5.3).

The Ghatal region is located in the eastern extremities of the KP command area. The Rupnarayan River—the regions' main drainage outlet (Fig. 5.4)—is fed by the water of both the DVC and KP, brought in by the southeast-flowing Dwarkeshwar River and the east-flowing Shilabati River, respectively. The release of excess discharges from the upstream reservoirs of the DVC and KP causes prolonged waterlogging in the Ghatal CD Block. Besides, due to tidal inflows in the Rupnarayan, flood spills could not drain out easily, hence worsening the situation. The low-lying area between the Damodar and Kangsabati Rivers is considered as an inter-sedimentary lobe pocket (Majumdar et al. 2010).

A number of tributaries and distributaries are connected with the Dwarkeshwar and Shilabati Rivers around the Ghatal CD Block (Fig. 5.4). Distributaries such as the Sankari and Old Rupnarayan join the Dwarkeshwar again downstream, above its confluence with the Shilabati. The Ketia *Khal*—a distributary running parallel to the Shilabati—joins the river again upstream of the Ghatal town. The Parang and Buriganga Rivers—tributaries of the Shilabati—contribute significantly to water flows during the monsoon months. A tiny channel also connects the Shilabati with the Old Kasai, a degraded distributary of the Kangsabati (Kasai) River. The Rupnarayan River, on the other hand, receives much water from a major distributary of the Damodar system—the Mundeshwari (Fig. 5.4).

3   Database and Methodology

The flood hazard susceptibility of the concerned region and the post-Phailin inundation scenario were analysed with the help of a digital elevation model (DEM) and multi-dated satellite images (Table 5.1). Various hydro-geomorphic techniques and spatial overlay followed by ground truth verification (GTV) were also used.

The SRTM DEM (v-3: 30 m) was primarily used to show the general physiography and slope. In conformity with the regional slope, the expected flow lines were generated in the GIS environment, following the extraction of flow direction and flow accumulation rasters. This DEM was also used for drawing an elevation profile so as to show the relative positions of flood causeways, the Shilabati River and circuit embankments. The satellite images by Landsat missions (false colour composites) and WorldView series (true colour composites) were used for visual image interpretation. On the basis of the data provided by the DoIW–GoWB (2014) and DFO (2014), vectors were generated so as to find out the flood susceptible areas and actual

Table 5.1 Details of the satellite images used in the study

Data particulars	Imaging date	Resolution (m)	Tile/scene identity	Data source
Shuttle Radar Topography Mission (SRTM) DEM	February 2000	30	v-3 (1 arc-second): n22-e87	United States Geological Survey (USGS)
Landsat-8 OLI	19 October 2013	30	Path-139, Row-44	United States Geological Survey (USGS)
Landsat-8 OLI	25 November 2013	30	Path-139, Row-44	United States Geological Survey (USGS)
Landsat 8 OLI	16 November 2014	30	Path-139, Row-44	United States Geological Survey (USGS)
Worldview-2	13 September 2009	1	–	Digital Globe (Google Earth)
Worldview-2	05 January 2010	1	–	Digital Globe (Google Earth)
GeoEye-1	18 April 2016	1.5	–	Digital Globe (Google Earth)

flooded areas during the post-Phailin event. In the field, a cross-sectional survey of the Chetua Circuit embankment was accomplished through the use of a transit theodolite. A Global Positioning System (GPS) receiver (accuracy: ±3 m) was used for identifying the positions of the flood embankments and inundation sites.

Oral interviews with administrative personnel and local citizens helped understand floodwater flow patterns. Gauge data were collected from the Office of the Assistant Engineer of the Ghatal Sub-division, the DoIW–GoWB. Information on geology, rainfall, flood history, tidal character, demography, etc. was obtained from different administrative departments, as well as from various reports and literatures.

4 Flood Characteristics in the Ghatal Region

From a hydro-geomorphic perspective, the Ghatal CD Block can be divided into three interfluvial zones, viz. (i) the Shilabati–Dwarkeshwar interfluve, (ii) the Dwarkeshwar–Old Rupnarayan interfluve and (iii) the Shilabati–Rupnarayan interfluve (Fig. 5.4). The Ghatal town is located within the last interfluve. Anything in excess of normal rainfall may cause flood-like situations in the Ghatal region. This area typically becomes flooded 3-5 times in a year (WBDMD 2014). Even during the British Raj, flooding in the Ghatal region was severe (Inglis 1909).

4.1 Circuit Embankments and Flood Causeways

In order to increase revenues for the erstwhile British Government, native landlords in the Ghatal region constructed circuit embankments in the floodplains so as to safeguard low-lying areas from flood spills and increase the amount of arable land (Fig. 5.5). Later on, in accordance with the 'Bengal Embankment Act' adopted in 1873, all the circuits and river banks in the area were acquired by the Government of West Bengal. As a result, the suspended sediment load brought in by the rivers could not spread into the floodplains and eventually became deposited in river beds. During the process said river beds were elevated in relation to their surrounding areas, causing an increase in flood levels every passing year. Not only were these circuit embankments built and positioned in a dysfunctional manner, they came to obstruct the natural outflow of spill water.

During the 1970s, the National Flood Commission recommended a complete overhaul of these circuit embankments. However, considering the socio-economic conditions of the area, no such action was taken immediately. Until the early years of the twentieth century, the Shilabati and its distributary Ketia *Khal* were capable of carrying excess discharges during flood-like situations. Gradually, channel siltation caused the reduction of these rivers' carrying capacity (WAPCOS 2009; Kar and Das 2014). As a result, the excess discharge carried within by these rivers breached circuit embankments on the left bank (Ghatal Circuit and Panna Circuit; Figs. 5.5 and 5.6)

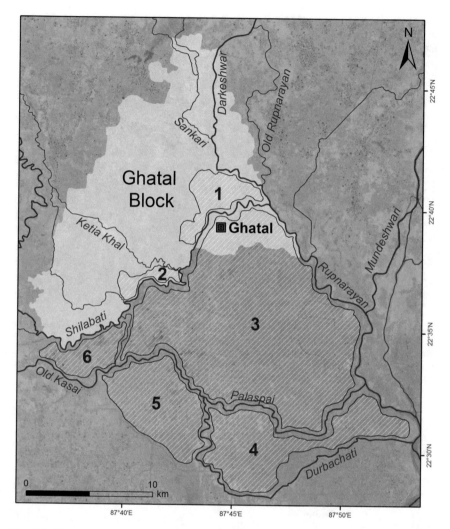

Fig. 5.5 Circuit embankments in and around the Ghatal CD Block. *Note:* The following circuit embankments are identified by numeric digits: **1** Ghatal Circuit (abandoned), **2** Panna Circuit (abandoned), **3** Chetua Circuit (functional), **4** Mohankhali Circuit (functional), **5** Dushaspur Circuit (functional) and **6** Narajal Circuit (abandoned). *Image* Landsat 8 OLI (Path-139, Row-44), 16.11.2014

many times, flooding the low-lying areas between the Dwarkeshwar and Shilabati Rivers in the process. Due to years of ineffectiveness, circuit embankments on the left bank of the Shilabati were declared abandoned by the erstwhile British Government in the 1940s. Repetitive incidents of flooding created natural flood water routes through these areas, locally referred as *Chatal* (the word refers to an extremely low and flat levelled surface). The DoIW–GoWB has identified these as flood causeways,

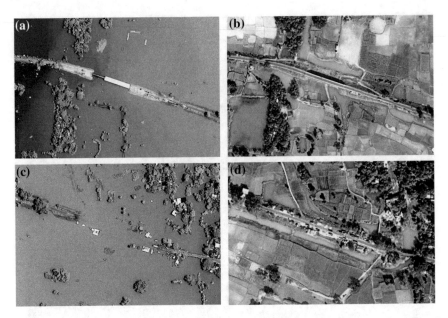

Fig. 5.6 'Flood causeways' in the Ghatal region during syn-flood (**a** and **c**) and post-flood (**b** and **d**). *Notes* A significant portion of the Ghatal–Chandrakona Road (State Highway no. 4) remains submerged during floods. Image **a** shows an under-construction flood flyover causeway, while image **b** shows the completed flyover construction. *Images* **a** and **c** Worldview-2, 13.09.2009. **b** and **d** Worldview-2, 05.01.2010

wherein the depth of flow is usually 3.5–4 m (Figs. 5.6 and 5.7). During floods, these causeways act as the main means of communication—for many of the roads in the Ghatal region become submerged.

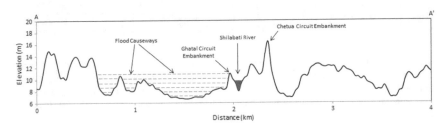

Fig. 5.7 Elevation profile (viewing north, vertical exaggeration: 60×) showing the relative positions of flood causeways in the Ghatal region, the Shilabati River and circuit embankments near Ghatal town. *Note* The Ghatal Circuit is currently abandoned; therefore, letting part of the spill water joins the Shilabati River. However, the Chetua Circuit protects the Ghatal town (located on the right bank of the Shilabati) from the flood spill

4.2   Effect of Tides

Tides, by triggering upstream flows through the Rupnarayan River, cause the stagnation of floodwater in the Ghatal region for longer periods of time. The Shilabati merges with the Dwarkeshwar at a place called Bandar to form the Rupnarayan. On the other hand, the Old Kasai, through its distributaries—the Palaspai and Durbachati—is connected to the Rupnarayan (Fig. 5.5). The connection between the Old Kasai and the Shilabati remains active during the monsoons. Through these connections, tidal discharges enter well into the Shilabati catchment. During spring tides, the tidal range is ~2 m at Bandar (WBDMD 2014). Flood impacts are at their worst if and when the outflow through the flood causeways of the Shilabati catchment area coincides with a spring tide inflow.

Besides tidal influence, significant amounts of discharge from the DVC command area carried by the Dwarkeshwar and Mundeshwari reaches the Rupnarayan River. Due to this, the Rupnarayan is often found to be in spate during high discharge events. This affects the Ghatal region upstream as floodwater from the Shilabati Basin struggles to exit through the Rupnarayan—the basin's only major outlet.

4.3   Extent of Flood in the Ghatal Region

Generally speaking, the Shilabati–Dwarkeshwar and the Dwarkeshwar–Old Rupnarayan interfluves are most affected by flooding due to both the basin-shaped topography of the region and the ineffectiveness of the circuit embankments on the left bank of the Shilabati (Figs. 5.6 and 5.11). However, flooding is not very common in the Shilabati–Rupnarayan interfluve itself, and this in spite of the erratic discharge outflows through the Rupnarayan. Standing at least 4 m higher than its surroundings and encircling the Ghatal town, the well-maintained Chetua Circuit embankment on the right bank of the Shilabati usually protects this interfluve from flooding (Figs. 5.7, 5.11 and 5.12).

Official records revealed that the Ghatal region was ravaged by devastating floods in 1888, 1913, 1942, 1956, 1959, 1968, 1973, 1978, 2000 and 2007 (DFEC 1943; WAPCOS 2011; DoIW–GoWB 2014; WBDMD 2014). In many occasions, the floodwater inundated an area larger than 100 km^2 for at least a month (Table 5.2).

5   Cyclone Phailin, October 2013

The October 2013 floods in the Ghatal region were entirely caused by torrential rains in the upper catchment of the Kangsabati–Shilabati Basin, following the landfall of the very severe cyclonic storm (VSCS) Phailin. With a maximum sustained surface wind speed of 220 kmph, it was the most intense tropical cyclone to reach landfall in

Table 5.2 Extent of the flooded areas in the Ghatal region and surroundings in relevant flood-prone years

Year	>30 days of inundation		15–30 days of inundation	
	Area (km^2)	Depth (m)	Area (km^2)	Depth (m)
1959	100	2.0	184	1.5
1967	100	2.5	69	1.5
1968	350	2.5	308	2.0
1973	208	3.0	150	1.5
1974	61	3.0	102	1.5
1975	104	2.5	110	1.5
1976	55	2.5	104	1.5
1977	100	3.5	130	1.5
1978	710	3.5	356	1.5
1999	78	3.0	100	1.0
2000	80	3.0	121	1.0
2007	233	3.0	400	2.0

Source WAPCOS (2011)

India since the 1999 Odisha cyclone (CWD–IMD 2013). As a tropical depression, the Phailin was formed within the Gulf of Thailand on around 4 October 2013. Over the next few days, it moved westwards and lay over the Andaman Sea. It eventually concentrated into a depression over the same region on 8 October. It further moved towards the west-northwest as an intensified, deep depression on 9 October morning and developed as a cyclonic storm. As it moved into the Bay of Bengal, the India Meteorological Department (IMD) named this cyclonic storm as 'Phailin' on the evening of 9 October. Moving towards the northwest, it further intensified into a severe cyclonic storm in the morning of 10 October and into a severe cyclonic storm in the forenoon of the same day over the east-central Bay of Bengal (CWD–IMD 2013). Phailin made its landfall on 12 October night near Gopalpur in Odisha. After the landfall, its wind speed reduced gradually as it further moved northward into the Chhota Nagpur Plateau region of Jharkhand state. On 13 October 2013, heavy rains of over 50 mm per day lashed the states of Odisha and Jharkhand. The western part of West Bengal, especially the West Medinipur District, received excess of 20% or more than the normal rainfall during the event (Fig. 5.8). The entire track of the cyclone is shown in Fig. 5.9.

5.1 Post-Phailin Floods in Ghatal

Due to the impact of the Phailin, the discharge of all the western tributaries of the Bhagirathi–Hooghly increased significantly (DoIW–GoWB 2014; WBDMD 2014). In the upstream of the Ghatal region, the Mukutmanipur (Kangsabati) Reservoir

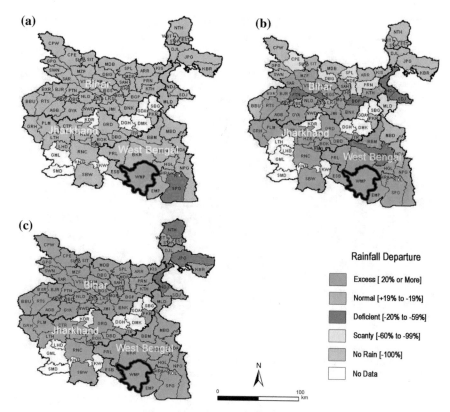

Fig. 5.8 District-wise rainfall departure of the eastern Indian states of Bihar, Jharkhand and West Bengal after the landfall of Phailin Cyclone in October 2013. *Note* The Chhota Nagpur Plateau region of Jharkhand and the western part of West Bengal received highly excessive rainfall within a short duration, which caused subsequent flooding in the Ghatal region. The West Medinipur District (highlighted with a red line) received in excess of 20% or more than the normal rainfall on 12.10.2013 (**a**), 13.10.2013 (**b**) and 14.10.2013 (**c**). After CWD–IMD (2013)

built on the Kasai and the Kadam Deuli Reservoir built on the Shilabati (Fig. 5.3) released plenty of discharge following the rainfall event. As a consequence, the low-lying Ghatal CD Block and its adjoining areas downstream of these rivers were submerged for two weeks (DoIW–GoWB 2014).

A full 93% of the 216 km^2 area constituting the Ghatal CD Block is susceptible to flood hazards; the post-Phailin outflow inundated an area of 161.7 km^2 (~75% of the Ghatal CD Block; Fig. 5.10). Widespread inundation occurred on the left bank of the Shilabati River, where the abandoned circuit embankments (Ghatal and Panna) were previously located. During the 2013 deluge, the Chetua Circuit sheltered a major part of the Ghatal town and surrounding areas from waterlogging (Figs. 5.11, 5.12 and 5.13). As the floodwater slowly drained into the Rupnarayan, the month-long stagnation of water occurred at a few places near the confluence of the Dwarkeshwar and the Shilabati (Fig. 5.11). The outflow draining out through the Rupnarayan was

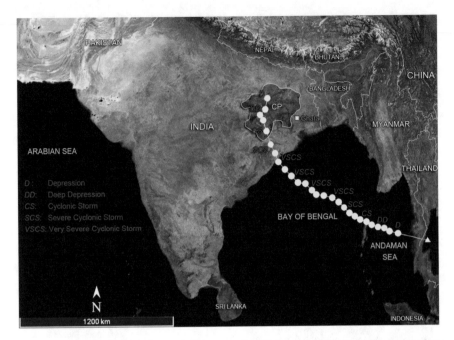

Fig. 5.9 Track map of the Phailin Cyclone between 8 and 14 October 2013. *Notes* The points show the location of the storm at 6-hour intervals. The arrow indicates the formation of the cyclone in the Gulf of Thailand. CP stands for Chhota Nagpur Plateau. The intensity levels transforming a depression into a very severe cyclonic storm were identified on the basis of the IMD classification scheme. *Tracking data courtesy* Joint Typhoon Warning Center (JTWC), The United States of America; *image courtesy* The National Aeronautics and Space Administration (NASA), The United States of America

joined by the discharge of the DVC command area brought in by the Mundeshwari, eventually flooding the downstream areas of the Rupnarayan catchment (DoIW–GoWB 2014).

The prolonged inundation in the Shilabati–Dwarkeshwar and Dwarkeshwar–Old Rupnarayan interfluve areas is evident from the gauge levels of the Shilabati and Rupnarayan in the Ghatal CD Block, as these two rivers seldom breached the danger level mark despite such a high-magnitude event (Fig. 5.14). A sudden increase in the gauge levels can be identified in both rivers on 15 October 2013, about 48 h after the occurrence of heavy rains in the Chhota Nagpur Plateau region.

5.2 Floodwater Routing

The floodwater routes of the post-Phailin flood correspond to the DEM-generated flow lines (Fig. 5.15). These flow lines are directed south-eastward and aligned across the Shilabati and Dwarkeshwar courses at a number of places, hence indicating the likely approach of floodwater towards these low areas. These flow lines specify the

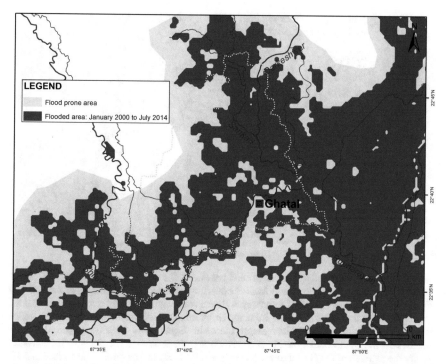

Fig. 5.10 Areas susceptible to flood hazard in the Ghatal region. *Note* The Ghatal CD Block boundary is demarcated by yellow dots. *Sources* Extent of flood-prone areas from DoIW–GoWB (2014); flooded areas in 2000 from DFO (2014)

inconsistency between natural slopes and the present courses of the rivers flowing through the region. This further puts in evidence the ill-structured construction of embankments and consequent floods.

6 Flood Prevention and the Ghatal Master Plan

At present, the maximum discharge carrying capacity of the Shilabati River is 650 m^3/s, which is only one-fifth of the maximum flood discharge observed in the Ghatal region in the last 50 years (WAPCOS 2011). Back in 1976, the Ghatal Master Plan (GMP) was introduced by the DoIW–GoWB so as to solve flood-related problems in the region. But it never really materialized in the three decades that followed.

In 2009, the federally owned Water and Power Consultancy Service (WAPCOS) was entrusted with the responsibility of designing a master plan that could be beneficial for both natural and cultural setups of the area. The WAPCOS-undertaken GMP

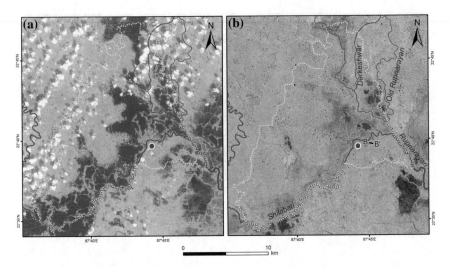

Fig. 5.11 'Phailin'-induced syn-flood (**a**) and post-flood (**b**) scenarios in the Ghatal CD Block (boundary demarcated by yellow dots). *Note* The interfluves between the Dwarkeshwar and Shilabati Rivers and Dwarkeshwar and Old Rupnarayan Rivers were significantly affected during the flood event. On the contrary, the Shilabati–Rupnarayan interfluve was least affected due to protection by the Chetua Circuit. Post-flood images indicate water stagnation for an extended period at a few patches near the confluence of the Shilabati and Dwarkeshwar Rivers. The red dots in the images specify the location of the Ghatal town. The extent B–B′ indicates the profile shown in Fig. 5.12. *Images* **a** Landsat-8 OLI (Path-139, Row-44), 19.10.2013. **b** Landsat-8 OLI (Path-139, Row-44), 25.11.2013

included an area of 1659 km^2, comprising 10 CD Blocks of the West Medinipur District and 3 CD Blocks of the East Medinipur District (WAPCOS 2009). Initially, the WAPCOS estimated 100 years probability of flood level (corresponding discharge of 3696 m^3/s) in the Ghatal region and planned to construct high earthen embankments on the banks of the Shilabati. Later on, it was realized that this could result into the Shilabati flowing ~5 m higher than the surroundings at bank-full stage and might lead to even more disastrous consequences than the usual floods in case of any accidental embankment breaching. Therefore, instead of 100 years probability, a 50 years probability of flood level (corresponding discharge of 3428 m^3/s) was estimated in order to control floods in the GMP command area (WAPCOS 2009). Considering the physiographic and cultural setups of the region, the WAPCOS advocated the formation of moderately high earthen embankments, along with regular de-siltation and widening of the river bed, thereby ruling out the possibility of any dam construction on the Shilabati course. It further proposed the installation of sluice gates and pumping stations for quickly draining out floodwater. A plan was also designed intended to re-erect the abandoned Ghatal Circuit in a systematic way (WAPCOS 2011).

In addition to these, a couple of flyovers were built on the Ghatal–Chandrakona State Highway no. 4 in the last few years, as this road repeatedly crosses the flood

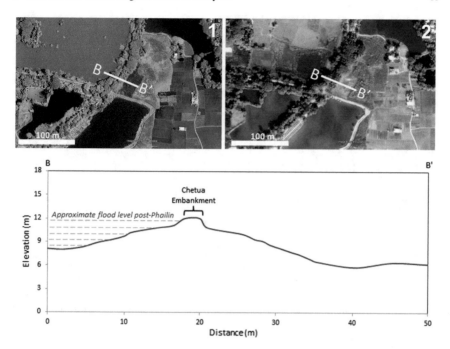

Fig. 5.12 Syn-flood (**a**) and post-flood (**b**) conditions near the Sandhur Bandh area of Ghatal town. *Note* The extent B–B′ indicates the profile across the Chetua Circuit embankment (viewing north, vertical exaggeration: 1.2×). The embankment protects the low-lying area on the eastern side during floods, as seen from the images and profile. *Images* **a** Worldview-2, 13.09.2009. **b** Worldview-2, 05.01.2010

causeways (Fig. 5.6). This road used to become submerged at a number of stretches for months—small country boats were used for communication between the disconnected segments. An inundated portion of the State Highway no. 4 during the post-Phailin flood can be seen in Fig. 5.16. In August 2016, the construction of the flyover over the nearest flood causeway of the Ghatal town was accomplished.

7 Conclusion

It is evident from this study that the unplanned and ill-thought constructions of the age-old circuit embankments are mainly responsible for the recurring misfortunes of the Ghatal region, including the post-Phailin floods. The embankments restricted the natural spills and winding of the rivers which shaped a low-lying pocket-lobe topography susceptible to inundation. Though the WAPCOS (2011) advocated various measures in the GMP, doubt persists regarding the outcome of if these plans would be implemented. *Firstly*, de-siltation and widening of the river bed as well as installation of sluice gates and pumping stations for quickly draining out the floodwater

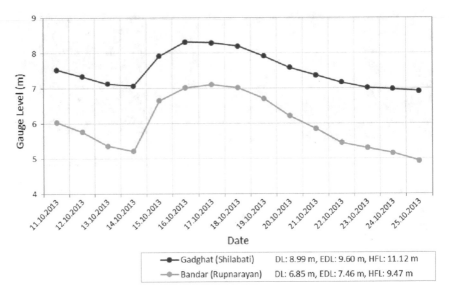

Fig. 5.13 Chetua Circuit embankment near the Sandhur Bandh area of Ghatal town. *Note* The extent B–B′ indicates the profile shown in Fig. 5.12. Photographed by the authors, 04.04.2014

Fig. 5.14 Gauge level fluctuations between 11 and 25 October 2013 at Gadghat (Shilabati River) and Bandar (Rupnarayan River) stations in Ghatal CD Block. *Note* In spite of the sudden increase in discharge and corresponding gauge levels, the Shilabati River was below the danger level mark throughout, while the Rupnarayan breached said danger level for three continuous days (16.10.2013–18.10.2013). Despite the flood situation in the interfluves, river discharge remained within the banks. *Data source* DoIW–GoWB (2014)

would not very be easy to accomplish as river courses are often not harmonious with the regional slope. *Secondly,* the erection of high, earthen embankments in accordance with the 50 years flood level is risky, as the river bed's present elevation of the is often greater than the adjoining region. In this situation, any instance of embankment breaching would create mayhem. *Thirdly,* for successful implementation of the GMP proposed by the WAPCOS (2011), permanent acquisition of 1407 ha land and temporary acquisition of 4180 ha land are required. In a densely populated region such as Ghatal, and considering the present eco-political excitements regarding land acquisition issues in West Bengal, this seems almost impossible. *Lastly,* the WAPCOS is in favour of constructing a high-level earthen embankment upstream of the Shilabati along its left bank. But this may lead to inundation in upstream areas at the

Fig. 5.15 Hydraulic routing of floodwater in the Ghatal region. *Note* The Ghatal CD Block boundary is demarcated by yellow dots. The rivers and DEM-generated flow lines are represented by turquoise and navy blue, respectively. The flow lines were generated from the 1 arc-second (30 m) SRTM DEM (v-3: Tile n22-e87) of February 2000. *Image* Landsat-8 OLI (Path-139, Row-44), 19.10.2013

Fig. 5.16 Nearest flood causeway of Ghatal town during syn-flood (**a**) and post-flood (**b**). *Note* The syn-flood photograph shows a submerged section of the Ghatal–Chandrakona Road (State Highway no. 4). *Photograph sources* **a** Obtained from Dolui and Ghosh (2013). **b** Photographed by the authors, 04.04.2014

lee side of the embankment the magnitude of which may surpass all previous inundation levels. The GMP may work out well in the short run, but uncertainty persists as regards its execution in the long run as it tries to solve the problem through hard engineering techniques alone. It must be kept in mind that such an approach during the British Raj is responsible for present flood-related issues in the Ghatal region.

Acknowledgements We express a deep sense of gratitude to Angshuman Adhikari (Sub-Divisional Officer, Ghatal, GoWB) and Namit Sarkar (Assistant Engineer, Ghatal Division, DoIW–GoWB) for providing us the required information. We also thank Probal Biswas and Saddam Mondal (ex-students of the Bhairab Ganguly College, Kolkata) for helping us during the field visit.

References

Agarwal RP, Mitra DS (1991) Palaeogeographic reconstruction of Bengal Delta during Quaternary period. In: Vaidyanathan R (ed) Quaternary deltas of India. Memoirs, Geological Society of India, vol 22, pp 13–24

Bagchi K (1944) The Ganges delta. Calcutta University, Calcutta, p 120

Bandyopadhyay S, Kar NS, Das S, Sen J (2014) River systems and water resources of West Bengal: a review. In: Vaidyanadhan R (ed) Rejuvenation of surface water resources of India: potential, problems and prospects. Geological Society of India, Special Publication 3, pp 63–84

Chandra S (2003) India: flood management—Damodar River Basin. The Associated Programme on Flood Management, World Meteorological Organization, 9pp

Chapman GP, Rudra K (2007) Water as foe, water as friend: lessons from Bengal's millennium flood. J South Asian Dev 2(1):19–49

CoI (2011) Census of India, Ministry of Home Affairs, Government of India. http://www.censusindia.gov.in/DigitalLibrary/Archive_home.aspx. Accessed 11 June 2015

CRED: Centre for Research on the Epidemiology of Disasters (2013) Flood. http://www.preventionweb.net/english/hazards/flood/. Accessed 06 Aug 2016

CWD–IMD: Cyclone Warning Division, India Meteorological Department (2013) Very severe cyclonic storm, PHAILIN over the Bay of Bengal (08–14 October 2013): a report, New Delhi, 42 p

DFEC: Report of the Damodar Flood Enquiry Committee (1943) Addendum. Reprinted in West Bengal District Gazetteers. Rivers Bengal 3(1):187–205

DFO: Dartmouth Flood Observatory (2014) Inundation image 080E030Nv3, 2 weeks ending 04.08.2014. http://floodobservatory.colorado.edu/Version3/080E030Nv3.html. Accessed 20 Aug 2014

DoIW–GoWB: Department of Irrigation and Waterways, Government of West Bengal (2014) Annual flood report for the year 2014. Directorate of Advance Planning, Project Evaluation & Monitoring Cell, Kolkata 117pp

Dolui G, Ghosh S (2013) Flood and its effects: a case study of Ghatal Block, Paschim Medinipur, West Bengal. Int J Sci Res 2(11):248–252

Duncan RA (1992) Indian Ocean, evolution in the hotspot reference frame. In: Nierenberg WA (ed) Encyclopaedia of earth system science, vol 2. Academic Press Inc., San Diego, pp 607–614

Fergusson J (1863) On recent changes in the delta of the Ganges. Q J Geol Soc Lond. Reprinted in: West Bengal District Gazetteers. Rivers Bengal 1:189–191, 218–222

GA: Geoscience Australia (2011) What causes floods? http://www.ga.gov.au/scientific-topics/hazards/flood/basics/causes. Accessed 08 June 2016

Goswami BN, Venugopal V, Sengupta D, Madhusoodanan MS, Xavier PK (2006) Increasing trend of extreme rain events over India in a warming environment. Science 314(5804):1442–1445

Inglis WA (1909) The canals and flood banks of Bengal. The Bengal Secretariat Book Depot. Reprinted in West Bengal District Gazetteers. Rivers Bengal 5(1):409–469

Kar NS, Das S (2014) বন্যাপ্রবণ ঘাটাল ব্লক - একটি ভৌগোলিক সমীক্ষা (A geographical analysis of flood vulnerability in Ghatal region). Bhugol o Paribesh 2(3):25–30

Lawver LA, Scalter JG, Meinke L (1985) Mesozoic and Cenozoic reconstructions of the South Atlantic. Tectonophysics 114:233–254

Majumdar M, Roy P, Majumdar A (2010) An introduction and current trends of Damodar and Rupnarayan River network. Impact Clim Change Nat Resour Manag 25:461–480

Niyogi D (1975) Quaternary geology of the coastal plain in the West Bengal and Orissa. Indian J Earth Sci 2(1):51–61

NDA: Natural Disasters Association (2013) What is flooding? http://www.n-d-a.org/flooding.php. Accessed 06 Aug 2016

Powell CM, Roots SR, Veevers JJ (1988) Pre-breakup continental extension in east Gondwanaland and the early opening of the eastern Indian ocean. Tectonophysics 155:261–283

Sanyal J, Carbonneau P, Densmore AL (2013) Hydraulic routing of extreme floods in a large ungauged river and the estimation of associated uncertainties: a case study of the Damodar river, India. Nat Hazards 66(2):1153–1177

WAPCOS: Water and Power Consultancy Services (India) Limited (2009) Executive summary, master plan and DPR for Ghatal Area, West Bengal. Irrigation and Waterways Directorate, West Bengal

WAPCOS: Water and Power Consultancy Services (India) Limited (2011) Executive summary, master plan and DPR for Ghatal Area, West Bengal. Irrigation and Waterways Directorate, West Bengal

WBDMD: West Bengal Disaster Management Department (2014) Flood. http://wbdmd.gov.in/Pages/Flood2.aspx. Accessed 13 Apr 2015

Part III
Water Trade-offs Between Sectoral and Regional Water Issues

Chapter 6
Social Capital and Irrigation Sustainability in Pakistan

M. Javed Sheikh and G. Mujtaba Khushk

1 Introduction

There is no substitute to water on this planet. Like water and life, Pakistan and agriculture are today inseparable. Pakistan cultivates three times more land than Russia! Agriculture contributes significantly to the GDP and engages around 43% of the national labour force (Nazish et al. 2013). However, water in the country has faced strong limitations (Abruzzi 1985), and its levels have been constantly declining (Hussain et al. 2004) and unpredictable. In addition, Pakistan is among the world's driest countries, with its lowest average rainfall below 240 mm a year. Pakistani farmers depend on an unreliable irrigation system to fulfil their basic needs and provide their livelihoods (John and Usman 2005). At the same time, population trends have led to growing water needs (Nakashima 2000).

Agriculture and water management are important tools for poverty reduction—they help provide both individual and collective benefits and enhance quality of life (Sheikh 2015). By contrast, water mismanagement further destabilizes irrigation infrastructure and leads to dysfunctional distribution and material decay, finally resulting in low crop production (Kulkarnir et al. 2011). This is why farming communities—especially in the Sindh Province of Pakistan—are largely dissatisfied with what they receive from the irrigation department (Sheikh et al. 2015). Simultaneously, interstate conflicts and local water disputes rarely gain international attention (Latif 2007). Hence, in order to counter such issues, local rural populations must turn to social capital so as to enhance agriculture-related benefits.

Generally speaking, farmers live in village communities (Kelsey 2013) sharing watercourses (Shivakoti 1991). Individual farmers have access to the community's

M. J. Sheikh (✉) · G. M. Khushk
Department of Rural Sociology, Sindh Agriculture University, Tandojam, Pakistan
e-mail: ruralsociologyst@gmail.com

G. M. Khushk
e-mail: mujtabaismail5@gmail.com

© Springer Nature Switzerland AG 2020
S. Bandyopadhyay et al. (eds.), *Water Management in South Asia*,
Contemporary South Asian Studies, https://doi.org/10.1007/978-3-030-35237-0_6

agricultural tools/implements and are expected to reciprocate by sharing irrigation water. Water reciprocity occurs when a farmer is in dire need of water but cannot access it because it is not technically "his turn." He asks community members for the needed water and promises to return the same amount of water when their turn comes. In rural Sindh, most farming communities have more or less informal relations with each other. They are free to regularly discuss irrigation and water issues so as to reach amicable conclusions. On the other hand, any directive from an outside community pertaining to water management is often not even considered (FAO 2004). Farming communities in the Sindh Province of Pakistan today tend to remain autonomous to say the least in the ways problems are identified and issues dealt with.

These communities perform a number of activities within the participatory irrigation management, including cleaning, lining and regularly renovating watercourses. Watercourse Associations also play a facilitating role between the irrigation department and farmers (Murray-Rust et al. 2000) in order to collect the Water Tax (Abyana), particularly during periods of water shortage. Concurrently, the absence of a drainage system/structure in a particular area may result in the total destruction of crops—chiefly at the lower elevated areas of agricultural land. This invariably (and understandably) leads to social conflict and strife (Sena and Michael 2006). Communities try to resolve such disputes peacefully and collectively so as to decrease the side effects of a flood situation (Howgate and Wendy 2009). However, some issues are not directly connected with irrigation management itself—for instance, crop theft, illegal grazing of animals, criminal activities, and issues relating to law and order. These issues are all best solved through increased participation and above all cooperation among farmers, preferably on the basis of mutual trust and mutual understanding (Bello 2005).

Carr et al. (2012) are of the strong opinion that farmer involvement in water management activities may increase levels of trust and reciprocity and lead to long-term, sustainable results. However, levels of community involvement, interdependence and reciprocity fluctuate both spatially and culturally (Sheikh et al. 2016b). Ultimately, the more the social capital, the more the reciprocity dynamics regarding water management—and the more the potential benefits. The exchange of irrigation timings contains perhaps the best reciprocal potential (Murgai et al. 2000). Once reciprocity relations are developed, benefits ensue. To a certain extent, water conflicts may all find easy solutions through farmer cooperation, as social bonds and norms are critical to sustainable development (Pretty 2003). Furthermore, farmers often consult each other about crop cultivations, laser levelling, poison, handling, application of fertilizers, reclamation of soil and so on. Reciprocity, a further extension of cooperation, may provide a golden chance for a community to develop a functional drainage structure, educational institutions, health services and security mechanisms. This in turn can lead to a row of economic, social and psychological benefits (Cohen and Uphoff 1977), as well as skill development (Hancock 2001), and environmental advantages (Reed 2008)—all of which could be categorized into individual and collective benefits (Cohen and Uphoff 1977). Social capital ensures sustainability.

2 Social Capital

Social capital is expressed in network structures, social resources (Seibert et al. 2001) and increased capacity for collective action (Esau 2008). It is "an instantiated informal norm that promotes cooperation between individuals." Social capital can be understood as a form of negotiation in social relations and associated projected returns (Berzina 2011). Indeed, social capital is one of the most difficult social phenomena to understand as it possesses multidimensional perspectives (Putnam, The Prosperous Community: Social Capital and Public Life, 1993). For that reason, no single theory is sufficient to understand it. Three scholars have contributed to a greater extent to the understanding and evaluation of social capital. Putnam (1993) highlighted bonding and bridging social capital; he was followed by Nahapiet and Ghoshal (Social Capital, Intellectual Capital, and the Organizational Advantage, 1998) who further categorized social capital into structural, relational and cognitive forms. In such scheme, structural social capital is defined as an overall pattern—morphology or network configurations—of connections between actors. Relational social capital concerns the personal relationships developed through face-to-face interactions within a group. Cognitive social capital refers to the ability by performers to build up mutually interpretive frameworks based on a shared language, codes and narratives. In this area, Woolcock (The Place of Social Capital in Understanding Social and Economic Outcomes, 2001) additionally linked social capital to the world of knowledge.

Putnam (Bowling Alone: America's Declining Social Capital, 1995) classified bonding social capital as the ways in which particular people within a uniform group or community engage in a closed set-up and express strong ties. By contrast, bridging social capital indicates refers to relations between different communities and networks whereas the members of one cluster have the rights to use the resources of another group through overlapping membership (Narayan and Cassidy 2001). Nahapiet and Ghoshal's concept of relational, social capital and Putman's concept of bonding social capital (Putnam 1993) are quite similar. Bonding social capital could be measured through "trust" levels, which are also a good indication of reciprocity levels among the members of a given community (Woolcock 1998). Later Woolcock (2001) divulged another form of social capital, which was termed linking social capital. Linking social capital is simply defined as the relations between individuals and groups within power-based relationships (Table 1).

2.1 Bonding Social Capital

Bonding social capital defines closed networks and depicts strong ties within uniform groups (Sheikh et al. 2015). For a deeper understanding, Putnam (1995) discussed some of the operating variables involved in bonding social capital. He was of the opinion that trust, norms, reciprocity, expectations and uniformity are all inherent to

Table 1 Selected social capital types

Dimension	Definition	Operational variables	Direction	Source
Structural	Overall patterns (morphology or network configurations) of connections between actors	Density, connectivity, hierarchy, solidarity	Horizontal (within a group)	Nahapiet and Ghoshal (1998)
Bonding	Involves closed networks and describes strong ties within homogeneous groups	Trust, norms, reciprocity, expectations, uniformity	Horizontal (within a group)	Putnam (1993)
Linking	Connections between individuals and groups in the hierarchy or power-based relationships	The extent of relationship with number of institutions and formal organizations	Vertical (individual to formal organization)	Woolcock (2001)

Source Sheikh et al. (2015)

bonding social capital. As this type of social capital lies within a given community, it is classified as horizontal. Panth (2010) holds that bonding social capital can be valuable for poorer sections of society to act collectively through groups and networks capable of articulating their common interests. The notion has been applied in the arrangement and capacity building among native groups, women's groups and rural communities. It is thus applicable to the existing situation of the WCAs in Sindh Province, Pakistan, and emphasizes in this context the general interests and collective power of inner-group connections in the effort to exercise collective action for common ends (Panth 2010). Within the concept of bonding social capital, "'trust" can be seen as and indeed has been seen by several scholars as the soul of relationships (Coleman 1988; Putnam 1993; Uzzi 1996; Snijders 1999)—and the concept can equally be seen as particularly helpful to understand reciprocity in water management. Trust is observed as the most imperative norm because it makes possible "the exchange of resources and information that are the key to high performance" (Uzzi 1996).

2.2 Structural Social Capital

Nahapiet and Ghoshal (1998) introduced the concept of structural social capital by defining it as an overall pattern of connection (morphology or network configurations) between actors. Structural social capital deals with individuals' behaviours and takes mostly the form of networks and associations (Uphoff 2001). Uphoff and Wijayaratna (2000) proposed that structural social capital is an asset to farming communities managing irrigation water. This form of social capital can be identified through the degrees of density, connectivity, hierarchy and solidarity characterizing social relations. It also acts within a given group or community (Nahapiet and Ghoshal 1998) and is thus considered horizontal as well. Structural social capital was defined as "group solidarity" by Coleman (1988), who posited great emphasis on network "closure" as based on robust interconnected social ties. Uphoff (2001) described it as the roles, rules, precedents and procedures ruling social interactions. Marsden and Oakley (1998) indicated that group solidarity is an often-ignored variable in studies regarding social capital.

2.3   Linking Social Capital

Woolcock (2001) identified networking as another form of social capital—the so-called linking social capital. Linking social capital refers to relations between individuals and groups within hierarchical or power-based relationships. Therefore, "networking" among farmers and formal organizations for example was to be evaluated according to the linking social capital model proposed by Woolcock (2001). Jennifer and Brian (2014) also defended the role of networks in both conceptual and empirical "links" between levels of analysis. Farmer associations' social networks play a significant role in learning—and in the adoption of new agricultural technologies (Mary et al. 2014) and related benefits (Rudd 2000). Linking social capital permits the build-up of resources, information and assets required for effective empowerment (Njuki et al. 2008). The authors of this chapter are of the strong opinion that linking social capital may be further subdivided into direct and indirect forms. In local cultural contexts, the role of landlords as intermediaries between farming communities and government and other agencies is crucial—as often only landlords have direct contacts with officials, while farmers mostly rely on indirect forms of communication.

Hence, the bottom-up progress of a given community is not as straightforward as Zimmerman and Warschausky (Empowerment Theory For Rehabilitation Research: Conceptual And Methodological Issues, 1998) implied in their theory of empowerment according to which "participation leads to empowerment." The fact is that the empowerment demands way more than simple participation in certain activities to achieve the set goals. Indeed, this "more than" can be seen as an indicator of effective social capital levels within a community (Sheikh et al. 2016b)—"effective" here meaning actually functioning as bonding dynamics capable of providing long-term, sustainable results (Sheikh et al. 2016a). Collective approaches encourage social capital production processes in terms of many different types of stakeholder networks because social capital is a multidimensional term (Jennifer and Brian 2014) and different cultures produce social capital differently (Dietlind and Hooghe 2003). Social capital may be expressed at various different levels and types at different geographical locations.

Social capital is also interrelated with other forms of capital (Berzina 2011) necessary to collective benefits. It offers greater innovation results and increased well-being within a community (Rudd 2000), leads to economic development (Giorgas 2007) and poverty eradication (Dietlind and Hooghe 2003), and has positive impacts even on academic achievements among students (Dufur et al. 2013). The benefits of social capital are also visible in business processes (Mačerinskienė and Aleknavičiūtė 2011)—it correlates positively with employment stability (Wu 2008) and is very useful in processes of conflict resolution (Marsden and Oakley 1998). Social capital can produce both tangible and intangible benefits in all human realms, including in so-called developed countries (Klein 2013). Consequently, a true application of social capital could align with other factors vital to development (Marsden and Oakley 1998) such as economics, sociology and management constituencies (Mohsenzadeh 2011).

Violence associated with water management is linked to a lack of social capital during a participatory process (Uphoff, Understanding Social Capital: Learning from the Analysis and Experience of Participation, 2001). Social capital has much to offer as regards addressing institutional obstacles linked to distribution inequalities and tackling the insecurity arising from the fact that water scarcity threatens livelihoods. Agricultural shortages cause increases in food prices, which much affect poor communities both rural and urban (Kugelman and Hathaway 2009).

Experience in other countries suggests that apart from the presence of effective irrigation and drainage systems it is also crucial to give farmers a decisive role in water management, for this will build confidence and empower farming (FAO Fa 2003). In order to bridge the gap in social capital dynamics, the authors attempted to review the overall theme of social capital so as to provide a clearer picture at the micro-level of farmers' ties in Sindh Province of Pakistan.

It was concluded in an article produced by Sheikh et al. (2016b) that social capital plays an important role in achieving collective goals such as agricultural benefits. In addition, researchers also found significant levels of social capital among the farming communities of the Sindh Province in Pakistan, based on reciprocity in the sharing of irrigation water. However, farmers severely lack networks of linking social capital, an issue which must be reviewed by both government authorities and policymakers. Although bonding and structural social capital are both reasonably present among local farmers, these too could be further enhanced and mobilized so as to bring harmony and progress in Sindh's rural economy.

3 Conclusion

The authors conclude that the "social capital" may play a decisive role in achieving rural development in Pakistan. Reciprocity in the sharing of irrigation water is present to a certain extent in rural communities; however, poor farmers lack sufficient linking social capital.

References

Abruzzi WS (1985) Water and community development in the Little Colorado River basin. Hum Ecol 13(2):241–269. http://www.jstor.org/stable/4602781

Bello DA (2005) The role of cooperative societies in economic development. Ahmadu Bello University, Zaria-Nigeria

Berzina K (2011) Enterprise related social capital: different levels of social capital accumulation. Econ Sociol 4(2):66–83

Carr G, Blöschl G, Loucks DP (2012) Evaluating participation in water resource management: a review. Water Resour Res 48(11). https://doi.org/10.1029/2011wr011662

Cohen J, Uphoff N (1977) Rural development participation: concepts and measures for project design, implementation and evaluatioon. Cornel University, Ithaka

Coleman JS (1988) Social capital in the creation of human capital. Am J Sociol 94:95

Dietlind S, Hooghe M (2003) The sources of social capital: generating social capital: civil society and institutions in comparative perspective. Bus Econ 288:19–27

Dufur MJ, Parcel TL, Troutman KP (2013) Does capital at home matter more than capital at school? Social capital effects on academic achievement. Res Soc Stratif Mobil 31(1):1–21

Esau MV (2008) Contextualizing social capital, citizen participation and poverty through an examination of the Ward Committee Systemin Bonteheuwel in the Western Cape, South Africa. J Dev Soc 24(3):355–380

FAO (2004) Tertiary level irrigation system management in the Chambal command by water user associations. United Nations, Rajasthan, India

FAO Fa (2003) Sindh water resources management—issues and options. United Nation, Rome

Giorgas D (2007) The significance of social capital for rural and regional communities. Rural Soc 17(3):206–214

Grootaert C, Tv Bastelaer (2002) The role of social capital in development: an empirical assessment, 1st edn. Cambridge University Press, Cambridge

Hancock T (2001) People, partnerships and human progress: building community capital. Health Promot Int 16(3):275–280

Homans G (1961) Social behaviour: its elementary forms. Harcourt Brace Jovanovich, New York

Howgate OR, Wendy K (2009) Community cooperation with natural flood management: a case study in the Scottish borders. AREA 41(3):329–340

Hussain I, Mudasser M, Hanjra MA, Amrasinghe U, Molden D (2004) Improving wheat productivity in Pakistan: econometric analysis using panel data from Chaj in the Upper Indua Basin. Water Energy Abstr 14(4):40–41

Jennifer WN, Brian DC (2014) Linking the levels: network and relational perspectives for community psychology. Am J Community Psychol 53(3–4):314–323

John B, Usman Q (2005) Water supply and sanitation. World Bank, Irrigation and Drainage, Hydropower, Water Resources Management, Pakistan. Oxford University Press, London

Kelsey M (2013) Communal approaches to water management. Retrieved from: http://systemsofexchange.org/communal-water-management/

Klein C (2013) Social capital or social cohesion: what matters for subjective well-being? Soc Indic Res 110(3):891–911

Kugelman M, Hathaway RM (2009) Running on empty: Pakistan's water crisis, 1st edn. Woodrow Wilson International Center, Washington, DC

Kulkarnir SS, Sinha SP, Belsare SS, Tejawat SC (2011) Participatory irrigation management in India: threats and opportunities. Water Energy Int 68(6):28–35

Latif M (2007) Spatial productivity along a canal irrigation system in Pakistan. Irrig Drain 56(5):509–521

Mačerinskienė I, Aleknavičiūtė G (2011) The evaluation of social capital benefits: enterprise level. Bus Manag Educ/Verslas, Vadyba ir Studijos 109–126. https://doi.org/10.3846/bme.2011.08

Marsden D, Oakley P (1998) Evaluating social development projects. Development Guidelines, No. 5

Mary T, Alexandra AB, Boris EB-U, Michée AL, David KO, EvelynNasambu O, Naveen P (2014) Effects of social network factors on information acquisition and adoption of improved groundnut varieties: the case of Uganda and Kenya. Agric Hum Values 31(3):339–353

Mohsenzadeh M (2011) Social capital influence on production. Interdiscip J Contemp Res Bus 3(2):497–510

Muhammad A (2011) Retrieved from AgriHunt: www.agrihunt.com/pak-agri-outlook/115-causes-of-low-yield-in-pakistan.html

Murgai R, Winters P, Sadoulet E, de Janvry A (2000) Localized and incomplete mutual insurance. University of New England, England

Murray-Rust DH, Lashari B, Memon Y (2000) Extended project on farmer managed irrigated agriculture under the National Drainage Program (NDP): water distribution equity in Sindh Province, Pakistan. International Water Management Institute (IWMI), Lahore

Nahapiet J, Ghoshal S (1998) Social capital, intellectual capital, and the organizational advantage. Acad Manag Rev 23(2):242–266

Nakashima M (2000) Water users' organization for international reform in Pakistan's irrigation sector. Faculty of International Studies. Hiroshima City University, Hiroshima, Japan

Narayan D, Cassidy MF (2001) A dimensional approach to measuring social capital: development and validation of a social capital inventory. Curr Sociol 49(2):59–102

Nasim A (2000) The use of irrigation systems for sustainable fish production in Pakistan. Pakistan Agricultural Research Council, Islamabad. Retrieved from http://www.fao.org/docrep/007/y5082e/y5082e05.htm

Nazish AR, Iqbal A, Ramzan M (2013) Impact of agriculture, manufacturing and service industry on the GDP growth of Pakistan. Interdiscip J Contemp Res Bus 5(4):727–734. Retrieved from http://journal-archieves35.webs.com/727-734.pdf

Njuki JM, Mapila MT, Zingore S, Delve R (2008) The dynamics of social capital in influencing use of soil management options in the Chinyanja Triangle of southern Africa. Ecol Soc 13(2):9

Panth S (2010) Bonding vs. bridging. The World Bank, Washington. Retrieved from http://blogs.worldbank.org/publicsphere/bonding-and-bridging

Pretty J (2003) Social capital and the collective management of resources. Science 302(5652):1912–1914. Retrieved from http://www.jstor.org/stable/3835714

Putnam RD (1993) The prosperous community: social capital and public life. Am Prospect 4

Putnam RD (1995) Bowling alone: America's declining social capital. J Democr 6:65–78

Reed MS (2008) Stakeholder participation for environmental management: a literature review. Biol Conserv 14(10):2417–2431

Rudd MA (2000) Live long and prosper: collective action, social capital and social vision. Ecol Econ 34(234):131–144

Seibert SE, Kraimer ML, Liden RC (2001) A social capital theory of career success. Acad Manag J 44(2):219–237

Sena L, Michael KW (2006) Disaster prevention and preparedness. Jimma University, Ethiopia

Sheikh MJ (2015) Farmers' participation, social capital and benefits in water management, Sindh Province of Pakistan. University Putra Malaysia, Selangor

Sheikh MJ, Redzuan M, Samah AA, Ahmad N (2015) Identifying sources of social capital among the farmers of the Rural Sindh Province of Pakistan. Agric Econ Czech 61(4):189–195. https://doi.org/10.17221/144/2014-AGRICECON

Sheikh MJ, Ma'rof R, Asnarulkhadi AS, Magsi H, Shahwani MA (2016a) Analysis of farmers' participation for water management in Sindh Province of Pakistan. Pak J Agric Agril. Eng Vet Sci 32(1):75–84

Sheikh MJ, Magsi H, Asnarulkhadi AS, Khushk GM (2016b) Dynamics of social capital among irrigation water users in rural Sindh Province of Pakistan. J Water Sustain 6(2):53–62

Shivakoti GP (1991) Organizational effectiveness of user and non-user controlled irrigation systems in Nepal. Michigan State University, Michigan

Snijders TA (1999) Prologue to the measurement of social capital. Tocqueville Rev 20:27–44

Thibault JW, Kelley HH (1952) The social psychology of groups. Wiley, New York

Uphoff N (2001) Understanding social capital: learning from the analysis and experience of participation. In: Ismail S, Partha D (eds) Social capital: a multifaceted perspective. The World Bank Publications, Washington, DC

Uphoff N, Wijayaratna CM (2000) Demonstrated benefits from social capital: the productivity of farmer organizations in Gal Oya, Sri Lanka. World Dev 28:1875–1890

Uzzi B (1996) The sources and consequences of embeddedness for the economic performance of organizations: the network effect. Am Sociol Rev 61(4):674–698

Woolcock M (1998) Social capital and economic development: toward a theoretical synthesis and policy framework. Theory Soc 27:151–208

Woolcock M (2001) The place of social capital in understanding social and economic outcomes. Isuma 2(1):11–17

Wu S (2008) Social capital in status attainment: structural constraints and individual determinants—a multilevel analysis. Duke University, ProQuest

Chapter 7
A Geosensor Site Suitability Analysis for Flood Early Warning/Management in the Indus River Basin

Mashal Riaz and Salman Atif

1 Introduction

The 2010 flood is considered as the worst in the history of Pakistan. It was caused by heavy rainfall during the monsoon season that hit Khyber Pakhtunkhwa (KPK) and northern Punjab and Sindh province. About 2000 people lost their lives and almost 2 million were otherwise affected. As regards infrastructure, 2 million homes, 400 miles of roads, 46 bridges and several railway lines were damaged; thousands of acres of agricultural land and wheat crops were destroyed (Akhtar 2011).

Flood management strategies in Pakistan mainly focus on flood warnings, measurement of water level in rivers, prediction of floods and rescue and relief work (Shafiq and Ahsan 2014). However, this strategy has several deficiencies. It gives no insight into short-term planning, prevention and preparedness. It only acts as a disaster response and lacks the incorporation of both environmental and sustainability concerns. National Disaster Management Authority (NDMA) in Pakistan managing the complete spectrum of disaster management in Pakistan on national level. The provincial and district disaster management authorities are responsible for practical implementation of disaster mitigation and management (Shafiq and Ahsan 2014).

The Pakistan Meteorological Department (PMD) is responsible for monitoring, weather updates and rainfall predictions in Pakistan. The organization has telemetric weather stations all over the country. Information is collected, communicated to and shared with the concerned departments for further dissemination. Flood Forecasting Division (FFD), Lahore, a specialized unit of the PMD, is mainly responsible for

M. Riaz (✉) · S. Atif
Institute of Geographical Information Systems, National University of Sciences and Technology, Islamabad, Pakistan
e-mail: mashal.riazz@gmail.com

S. Atif
e-mail: salman@igis.nust.edu.pk

© Springer Nature Switzerland AG 2020
S. Bandyopadhyay et al. (eds.), *Water Management in South Asia*, Contemporary South Asian Studies, https://doi.org/10.1007/978-3-030-35237-0_7

flood forecasting, river stream flow forecasting and water management at dams, especially during the monsoon (Pakistan Meteorological Department 2016). The PMD shares weather and flood advice with the National Disaster Management Authority (NDMA), concerned federal ministries and departments as well as the military in order to articulate further action and preparation (Pakistan Meteorological Department 2016). NDMA disseminate this information to Provincial Disaster Management Authorities (PDMA) and District Disaster Management Authorities (DDMA). The DDMAs convey the flood updates to the Tehsil Municipal Administrations (TMA), the Union Councils (UC) and ultimately to the concerned communities. Tehsil is the sub-division of a district and in turn is divided into a number of other smaller units called Union Councils (Population Association of Pakistan 2016). These administrative authorities at UC level are responsible for dissemination of primary information to the communities (Fig. 1).

The weather monitoring network consists of a 10 cm Doppler radar installed at the Federal Flood Division in Lahore. The Doppler radar in Lahore has a range of 450 km (Pakistan Meteorological Department 2016). The most essential element for quantitative flood forecast is the hydrological data that is provided by the Water and Power Development Authority (WAPDA) and the Provincial Irrigation Department. The Provincial Irrigation Department is mainly responsible for managing small nullahs. It observes, measures and reports altering water levels in these nullahs and their possible impact on the occurrence of a flood. The telemetric stations installed for the Indus River System Authority (IRSA) were mainly mounted by the WAPDA; they collect hydrological data used for flood forecasting (Pakistan Meteorological Department 2016).

The flood monitoring and early warning systems adopted globally are based on real-time data collection instead (Qi and Altinakar 2011). Geosensors are used to

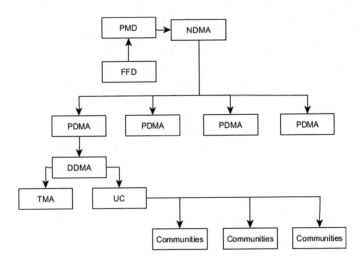

Fig. 1 Chain of information flow during a disaster in Pakistan. *Note* Information is handled initially between the PMD and FFD. It is then sent to other bodies via the disaster management authorities

collect data on river flow, precipitation, wind direction, soil moisture, humidity, etc. in real time and analyzed readily. These systems are free of human errors and the data acquisition can be accomplished at any time. Wireless sensor networks are also efficient in transmitting messages to greater distances and have the capability of sensing as well as receiving information (Sunkpho and Ootamakorn 2011). These sensor networks are cost-effective, using readily available energy sources and have the capability of operating even if one of the sensors malfunctions. Pakistan is in need of a similar system wherein flood monitoring and early warning information dissemination is based on real-time data collection and analysis.

As a general trend within ongoing innovations in the Internet of Things (IoT) arena, a sensor based on a real-time monitoring approach for floods and similar disasters seems to take shape. Having started with satellite remote sensors, the approach in the last decade is more based on ground-based sensors (Hong et al. 2007; Lee et al. 2008).

2 The Indus River Basin

Flood risk assessment is used for floodplain management, disaster evacuation, warning and evaluation as well as improving the public's awareness of flood risks. Floods often take place in the vicinity of rivers and plains, indicating a higher risk in such areas (Basha and Rus 2007). Remote sensing and statistical information are used to present these hazard factors (Jiang et al. 2008). Remote sensing is particularly important, for it provides satellite imagery for generating hazard and risk maps. These maps are then utilized for determining the extent of damage and articulating future planning (Zlatanova and Fabbri 2009).

The Indus Basin Irrigation System (IBIS) is one of the largest irrigation networks in the world. Its system of irrigation is supported by three storage reservoirs. There are 19 headworks and barrages with 12 link canals constructed on the Indus River. The most prominent such barrages constructed are the Chashma, Guddu, Jinnah, Kotri, Sukkur and Taunsa barrages. The area suitable for cultivation in the Indus River Basin is about 34.5 million acres (Ahmad and Majeed 2001). Out of this, 12.25 million acres are irrigated by the Mangla and 22.25 million acres by the Tarbela reservoirs.

The variation in river flow is an issue that causes irrigation water supply problems during the Rabi season, when the flow is lower, and during the early Kharif season (Dorosh et al. 2010). Wheat, sugarcane and cotton are the most important crops grown in the Indus River Basin and the lands they grow on are irrigated by the Indus River. The large number of people residing in the vicinity of the Indus River and the agricultural lands present there put them in a direct line of risk in case of flood.

The agricultural lands too are prone to adverse effects by floods due to their location; the damage to crops results in food shortages once the flood water has drained. There is a need for a robust flood early warning system, proper mitigation techniques and management practices, to minimize the devastation caused by floods—affecting populations, agricultural lands and infrastructure, thus straining the country's economy. The early warning system needs to be upgraded using real time technology in order to enhance the flood management practices otherwise the damage caused by floods in Pakistan may continue to increase.

The main reason for occurrence of floods in the country is the monsoon rainfall, which occurs every year due to a shift in wind direction (Aon Benfield 2014). There are typically two cycles of monsoon, one in summer and one in winter. The summer rainfall generally enters Pakistan from the east and north, and an ample amount of rainfall is received almost every year in the northern and north-eastern parts of Pakistan (Salma et al. 2012).

3   Flood Protection Plans

Major floods also occurred in the year 1950, 1956 and 1976. At that time, there was no flood management plans or policies in place intended to prevent such disasters. In 1973, a major flood hit Pakistan killing 474 people and caused damage to livestock, infrastructure and agriculture worth 160 billion PKR (Federal Flood Commission 2009). After this devastating event, flood management in Pakistan began as a three-levelled plan (Tariq and Van de Giesen 2012).

Flood management practices and techniques employed in Pakistan also consist of constructing flood protection structures such as spurs, studs and embankments. These protective measures were initially undertaken by provincial governments acting to solve local flood issues (Baig 2008). The Federal Flood Commission was later established in 1977 so as to better manage floods in Pakistan. A National Flood Plan was articulated in 1978. Three National Flood Protection Plans—from 1978 to 1987 (NFPP-I), 1988 to 1997 (NFPP-II) and 1998 to 2007 (NFPP-III)—have been created and executed. Up to 2009, an estimated amount of PKR 17.8 billion has been spent by the government—rescue and relief costs not included (Federal Flood Commission 2009). About 1350 protection schemes against floods were completed in the NFPP-I wherein flood protective structures were either constructed or repaired. The total amount spent on NFPP-I schemes was PKR 1.73 billion (Shaikh 2008).

Flood management can be considered as a spatial problem (Simonovic 2002; Luino et al. 2009). Flood protection is possible to achieve by undertaking structural measures, e.g. dykes, diversion channels, reservoirs, dams, etc. Non-structural measures include flood warnings, mass evacuation, etc. Spatial representation of the consequences of these flood mitigation techniques is a better insight into their effectiveness and aid in decision-making processes (Jonkman and Vrijling 2008). Comparing various flood protection techniques and evaluation of their impacts are based on a number of different criteria. These criteria usually include loss of human life and urban flood damage (Bouwer et al. 2010; Pingel and Watkins 2009).

Due to the changing dynamics of global climate, many urban areas all over the world are at a risk of flooding due to the rise in sea levels and climate instability. Extreme weather conditions that trigger floods include heavy rainfall and flash floods. Managing flood risks is important since it aids in avoiding economic, human and infrastructure-related losses. The routes for transportation and property can be saved by adopting a proactive approach towards flood early warning and management (Libelium 2016). Flood prediction has often been undertaken through manual readings or via satellite imagery; however, these methods may not always prove to be the best available options. Manual readings refer to physical recordings of the rise in water levels close to rivers. In developing countries, satellite data may be inadequate and available only between extensive time intervals. Therefore, wireless sensor networks can be considered an economical and accessible substitute that can be used for detection of early signs of floods, flood predictions and for monitoring flood-affected areas. The motes can be placed along the course of the river in order to measure water levels. Flood alerts can be generated wirelessly via SMS or by posting on an internet database (Libelium 2016). The geosensor network can be used for monitoring weather conditions, rainfall measurement and precise flood forecasts. It can also be used for determining the possibility of floods in coastal areas and rivers.

In a flood warning system, rainfall and water flow level-related data are measured at various points in the catchment area. The results of a warning system are aimed at an autonomous monitoring system for collecting data on water levels in several different locations during floods (Basha et al. 2008). Various sensor network systems are designed for monitoring. Of greater relevance is the work on environment monitoring.

4 Real-Time Data and Flood Monitoring

Location-based services (LBS), participatory sensing (PS) and human-centric sensing (HCS) are some of the applications used in the search for various warning applications. Participatory sensors are used to collect data so as to measure and monitor any variable of interest in a hazard-prone community, city, country or even on a global scale. Some participatory sensors are applications that collect air quality measurements in real time for determining the air pollution index. GPS locations are used to assess traffic congestion and travel times, noise, humidity and temperature measurements so as to create real-time maps. Human-centric sensing applications are used to collect health-related data of a user or group of users. Participatory sensors and human-centric sensors both integrate specific sensors into cellular telephones for measuring the required variables. Examples of variables include carbon dioxide, temperature, humidity, heart rate, breath depth, oxygen in the blood, etc. that need specialized sensors other than the ones available.

A comprehensive review of numerical weather predictions, also known as ensemble prediction systems (EPS), has been devised. The operational and pre-operational flood forecasting systems are designed based on a detailed analysis of the catchment area and the time period of the disaster (Cloke and Pappenberger 2009). The key challenges in the research study are the quantitative and evidence-based case studies for false alarms and future rainfall-related uncertainty. Research has been carried out regarding weather forecasting based on distributed hydrological modelling. The research is based on a metrological database spanning the last 30 years named COSMO. It is used to predict peak discharge climatology for flash flood events (Alfieri et al. 2013). A statistical model for prediction of floods using ad-hoc wireless sensor networks (WSN) has been introduced. Multivariable regression techniques have been used for calculating the extreme flow of flood waters.

More recently, it has become evident that sensor technology on the ground is a key to information reliability in disaster prevention. While satellites for earth imaging have been widely used for post-disaster information gathering, in situations arising both before and after disasters, more invasive means are needed.

Sensor nodes are being used to this end; for example, possible volcanic activities (Werner-Allen et al. 2005).

5 Data Processing and Site Identification

MODIS 09GQ data was used for recurrence mapping so as to identify the most vulnerable areas with respect to flood hazards. MODIS Surface Reflectance products are better suited for flood mapping because they provide an estimate of surface spectral reflectance as measured at any point on ground level (Nigro et al. 2014). MOD09GQ provides channels 1–2 at 250-m resolution on a daily basis. The main advantage of MODIS data is that it has a near-global spatial (250–1000 m) and temporal resolution (imagery captured one or two times a day). It is a gridded level-2 product in the Sinusoidal projection (Nigro et al. 2014). Moreover, all historical satellite data captured from 2000 onwards is made available (Park and Kwak 2011).

Band 1(Red) and 2(NIR) of MOD09GQ are 250 m resolution adequate for flood mapping on a wide area. Resolution of MODIS data does not provide the flexibility of mapping small and narrow features such as constricted river networks. However, the consistency shown by spatial and temporal resolution is ideal for detecting water movements (Ticehurst et al. 2014). For mapping flood plain areas, MODIS data for 2003, 2005, 2009, 2010, 2011, 2012, 2014 and 2015 was used. It focused on the areas that were affected by floods in the last 15 years—generally speaking, the entire stretch of the Indus basin (districts on either bank of the Indus River and its tributaries are regularly affected). The step-wise methodology shows the processes used (Fig. 2).

The first step was the MODIS data acquisition which was downloaded from the LAADSWEB. The Bands of 250 m resolution were extracted of 250 m resolution were extracted by processing the raw data in the MRT Tool.

Bands 1 and 2 were combined to obtain an NDVI. The inundated areas were extracted by applying a water threshold of 0.4–0.7, which is the used range for extracting water pixels in MODIS. After density slicing, the raster colour slices were exported both in the form of raster and polygon shapefiles for the preparation of flood recurrence maps.

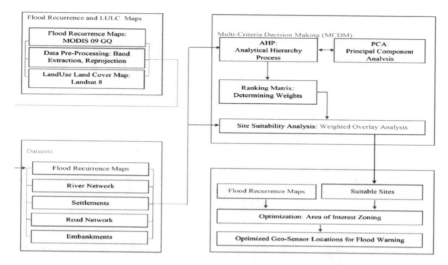

Fig. 2 Methodological framework for optimized geosensor locations for flood early warning

The raster colour slices were processed and reclassified via the duration tool. Events of interest are coded as 1 and non-events as 0 (Dilts 2015). The result of this process were output rasters which were assigned appropriate colours to show the inundation.

Five datasets were utilized in the weighted overlay analysis model. These included recurrence, roads, networks, rivers, settlements and embankments. These factors play an important role in any decision-making process relevant to flood management or early warning. Settlements and citizens are the most vulnerable factors when it comes to flood management. Roads play an important role in the rescue process and also in the deployment and maintenance of geosensors. Embankments are essential for preventing flood waters from spreading towards settlements and causing devastation.

The input weights for the parameters used were calculated using the analytical hierarchy process (AHP) (Mishra et al. 2015). In this technique, the inputs were compared in a pairwise manner. These criteria were then compared in a pairs wherein weights were assigned to them. The weights lie typically in a range from 1 to 9 (Bunruamkaew and Murayam 2011).

Principal component analysis was performed after the AHP in order to determine whether the weights assigned to inputs were consistent. It served as a technique to define the relative importance of each parameter in the weighted overlay analysis and their impact on the final result of weighted overlay analysis.

The most important step was the weighted overlay analysis for determining the most suited locations. The weights calculated by AHP were used for the execution of the weighted overlay analysis model, the execution of which generated the final result (Fig. 3).

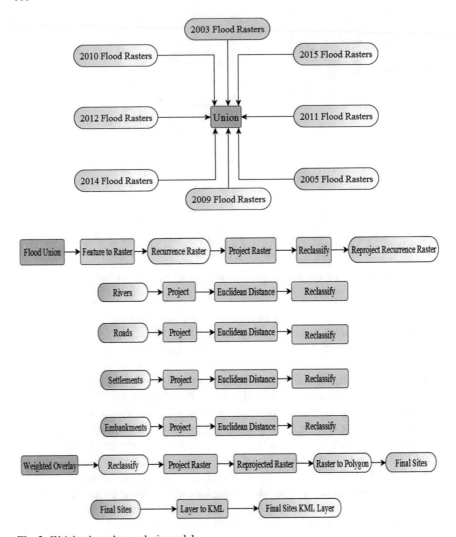

Fig. 3 Weighted overlay analysis model

6 Results and Discussion

Flood recurrence maps for the years 2003, 2005, 2009, 2010, 2011, 2012, 2014 and 2015 were created that identified areas repeatedly inundated by floods. The flood duration maps revealed that most areas which suffered losses are located on the lower Indus Basin, particularly southern Punjab and Sindh.

The 2003 flood had a devastating impact on southern and south-western parts of Pakistan. The first round of monsoon started off in mid-July, causing rivers to inundate. This spell lasted until the end of the month. Heavy rainfall ceased after

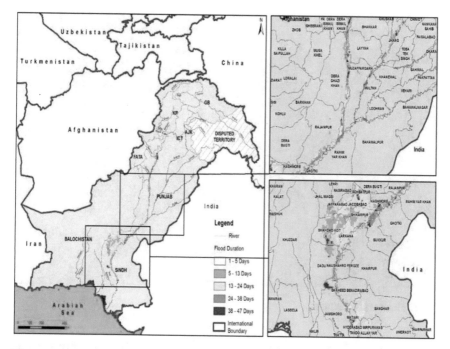

Fig. 4 Flood extent for 2003

two weeks but rainfall of low intensity continued at scattered locations in Sindh. The majority of the roads were severely damaged. The districts most severely devastated included Thatta, Larkana, Hyderabad, Dadu, Badin and Nawabshah. The generated flood map (Fig. 4) shows the affected areas. Some of the districts of southern Punjab such as Rajanpur, Dera Ghazi Khan, Muzaffargarh and Multan were also affected by heavy rainfall—but not as severely as districts in Sindh such as Kashmore, Jacoba-bad, Shikarpur, Larkana, Hyderabad and Tando Allah Yar. Although there is a line of embankment present that prevents flood water from spreading, in the case of 2003 floods the protective structures breached, causing the flood water to spread and stay stagnant in different areas of Punjab and Sindh for a few weeks. Some of the affected plain area districts such as Kashmore, Jacobabad, Shikarpur, and Mirpur Khas were flooded for about 13–24 days. A few areas in these districts namely Multan, Muzaf-fargarh, and Shaheed Benazirabad were severely devastated and lay flooded for over 40 days. Another factor that increases flood damage is the presence of barrages. The most highly inundated areas were located near the Sukkur, Lloyd or Guddu barrages. It has been a common practice to release excess water from barrages during flood when the water level crosses the maximum holding capacity of these reservoirs. This often leads to an increase in the magnitude of the flood.

The 2005 flood was different from the others as a massive heatwave in south-ern Asia in June–July 2005 caused the mountain ice to melt rapidly, which in turn inundated the rivers in the region. In Pakistan, the glacial melting resulted in a huge

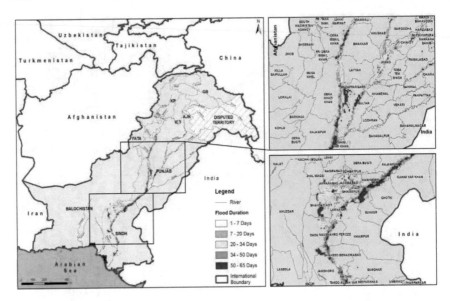

Fig. 5 Flood extent for 2004

volume of water flowing into the rivers, particularly the Indus. Water levels continued to fluctuate throughout the month of July resulting in flooding in the Khyber
Pakhtunkhwa, Punjab and Sindh provinces. The flood map (Fig. 5) shows that areas
located near rivers—especially, the Indus River—were severely damaged during the
2005 floods.

The 2009 flood mostly affected certain parts of Sindh (Fig. 6). Spells of heavy
rainfall were recorded majorly in Sindh while Punjab and Khyber Pakhtunkhwa were

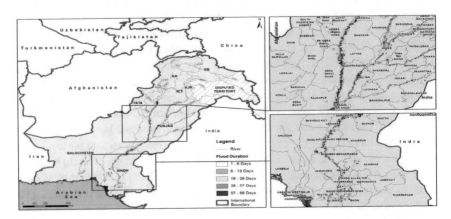

Fig. 6 Flood extent for 2009

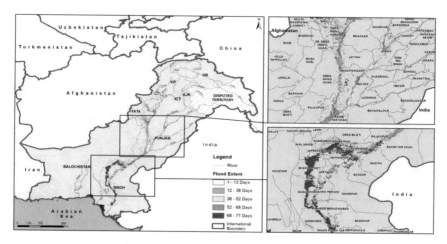

Fig. 7 Flood extent for 2010

affected in a few areas. The major priority zones in terms of flood were found to be mostly located in Sindh.

The 2010 flood is considered among the most devastating flood events recorded in modern history. Abnormally heavy rainfall was recorded in the monsoon season in 2010 causing water levels in rivers to rise exponentially. Starting from mid-July, the rainfall continued for several days, affecting large areas of the country. The flood affected the northern region of Pakistan; towards mid-August, it had reached Sindh via Punjab, causing damage to human lives, infrastructure and agriculture. This abnormally high rate of flood and devastation is evident from (Fig. 7). Most sites identified as high priority zones were located in Khyber Pakhtunkhwa, Punjab and Sindh. These locations have been identified as vulnerable locations in almost all the flood events that have occurred in Pakistan.

The 2011 flood was caused by unusually high rainfall in South Asia in general. Pakistan experienced widespread rainfall, particularly in the Sindh province where the variation in precipitation went from 400 mm to a value slightly higher than 1000 mm. The heavy downpour continued from 1 July to 28 September 2011. The Thar Desert was highly devastated, since soil infiltration rates were high and caused minimum water runoff. While rainfall was experienced in many parts of the country, it was Sindh that had the maximum impact of these rainfall incidents (Fig. 8). For several days, the flood-affected areas remained inundated, causing heavy damage to human life, property and infrastructure.

The 2012 flood was predicted by the PMD—keeping in view the previous two flood events in 2010 and 2011. From the beginning of July to mid-August, monsoon rainfall was scant in Pakistan. In the last ten days of August however, heavy rainfall caused rivers to inundate (Fig. 9). This spell of rainfall continued into September and caused devastation in the Sindh province. The 2012 flood was less intense when compared to the previous events in 2010 and 2011; however, it managed to affect several districts for a number of days.

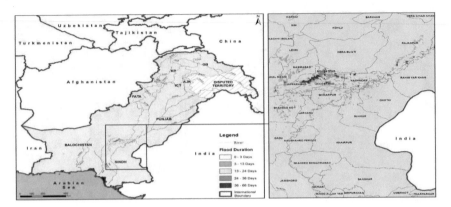

Fig. 8 Flood extent for the year 2011

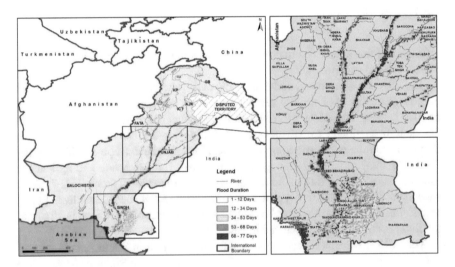

Fig. 9 Flood extent for the year 2012

The 2014 flood was caused by heavy rainfall in the first week of September which caused water levels to rise in Jhelum, Chenab, Ravi and Sutlej. This caused flooding in some parts of Punjab and Sindh. The 2014 flood was more concentrated around the river; flood waters did not spread to far-off areas (Fig. 10). Owing to rainfall that lasted the entire monsoon season, the riverside population was negatively affected.

The 2015 flood was caused by both mountain ice melting and torrential rains in July. As a result of this, two major flood waves were generated in the Indus River in the southern part of Punjab. The rainfall caused flash floods in some southern Punjab districts (Fig. 11).

The pairwise comparison led to the determination of final weights via four iterations. The iterations are done in order to get the most consistent values. As a result

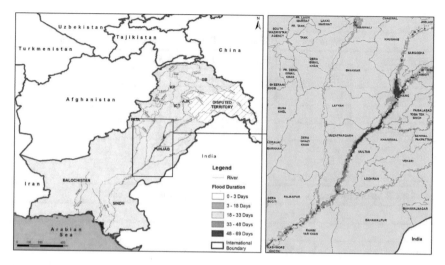

Fig. 10 Flood extent of 2014

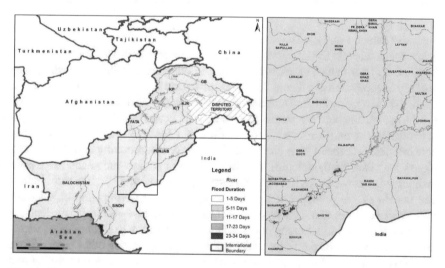

Fig. 11 Flood extent for 2015

of this process, a consistency index is determined which is further used for the calculation of individual percentages or inputs weightage in the normalized matrix. The percentage of flood recurrence in the final output is determined to be 56.9%, which means that it has the highest impact on the determination of suitable sites for geosensor networks. The rivers have an impact of 14.7%, which reveals their influence on the final result as floods mainly originate from them. Road networks have a weightage of 6.3% while settlements and embankments have an influence weight of 10.3 and 11.6% on the final site determination as a result of weighted overlay

Table 1 Criteria weights assignment

	Recurrence	Rivers	Roads	Settlements	Embankments
Recurrence	1	5	7	5	5
Rivers	1/5	1	4	2	2
Roads	1/7	1/4	1	1/3	1/2
Settlements	1/5	1/2	3	1	1/2
Embankments	1/5	1/2	2	2	1

Table 2 Normalized matrix for criteria weights assignment

Analytical hierarchy process		Consistency check
0.569	56.9%	6%
0.147	14.7%	
0.065	6.5%	
0.103	10.3%	
0.116	11.6%	

analysis (Tables 1 and 2). The results show that the areas affected by floods are the prime focus wherein the sensor network ought to be installed. The sensors are often installed on river banks, dams, barrages or any other suitable locations where they are easy to access and can record their respective natural phenomenon with accuracy.

Statistical approaches and correlation analysis have often been used for determining the correlation between factors that have an impact on the design of hydrological networks (Moss and Karlinger 1974; Benson and Matalas 1967). Over the years, network design theories have continuously evolved, incorporating correlation techniques into their methodologies. In order to determine how pragmatic the weight assignment exercise in AHP was, a PCA was performed (Table 3). The most important benefit of the principal component analysis is its determination of the most influential variables and their respective impact on the final result. It is calculated based on the variability among components. The first principal component, recurrence, was found to be the factor with highest weightage among other inputs. Since the damage incurred by floods is the most important aspect of flood warning and mitigation, it is understandable that this variable is the most important factor in site suitability analysis. The second principal component, the rivers and road networks, shares almost equal importance in terms of contribution to flood management. River

Table 3 Principal component analysis

	PC1	PC2	PC3	PC4	PC5
Variance	4.77	0.12	0.08	0.03	0.0
Proportion	0.95	0.02	0.02	0.01	0.0
Cumulative proportion (%)	95.5	97.8	99.4	100	100

networks owe their importance to the fact that maximum damage is caused in the areas that lie near river banks. Road networks are mainly significant for their role in flood management activities such as rescue and response. In the case of geosensor networks, they have an added role of enabling access to areas where the sensors have to be installed and their subsequent management. The factors with the least proportion or impact on results were determined to be the settlements and embankments. Flood warning elements are installed near such locations where they are able to perform well and provide timely information and warning. Embankments have proved to be an anomaly in the case of flood events. The reason being that the existing embankments have been deformed—either naturally due to high water pressure during floods or purposely so as to divert the path of water from its natural course.

For weighted overlay analysis the datasets are preferably created in raster format. In this case, the model was so devised that the vector datasets were converted to raster during execution. All the recurrence layers for the given years were merged in order to get a single raster indicating the areas inundated—(Ali 2013). Based on the criteria weights calculated via the analytical hierarchy process, weights were assigned to the map layers. The final output gave the most suitable sites for installation of geosensor networks in the Indus River Basin. The final sites obtained were converted into a KML layer.

Weighted overlay analysis is significant in the sense that it aids in the application of a common scale of values into inputs that are diverse and dissimilar in nature. This technique provides an integrated analysis. Input parameters that have different scales of importance and values are easily combined to provide the desired result.

The most prominent locations identified were mostly situated in Punjab and Sindh (Fig. 13). The main locations repeatedly inundated by floods were identified as the most suitable sites for the installation of geosensors. The final locations identified were in polygon format were in polygon format which were converted into points.

A geosensor network not only receives but also transmits information. A total of 3373 points were obtained by converting the polygons into points. A geosensor network needs to be optimal since a high-density network often faces communication issues in both sending and retrieving information (Basha 2010). Out of the resultant locations, 100 points covering the entire study area were chosen (Fig. 14). The sensor nodes of a WSNS are positioned relative to one another. For positioning a node, anchor nodes—also known as beacon nodes or relay stations—are set in the network. The scale of anchor nodes within the network is smaller in comparison with the sensor nodes; however, they are used as reference points for determining the position of unknown nodes. The same process is implementable on proposed geosensor sites, wherein some of the nodes identified will be used as beacon nodes.

The nodes are either pre-calibrated via a network routing algorithm or carry a positioning equipment such as GPS (Zhongguo and Tianfang 2015). Routing algorithm is an information-driven sensor querying or a multi-path routing algorithm that can also be applied. Solar panels are a good and reliable source of power for geosensor nodes. They can be mounted at the time of installation of geosensors for stable power supply.

Warning
A warning is issued when a hazardous weather or hydrologic event is occurring, imminent or likely. A warning means that weather conditions pose a threat to life or property. People in a storm's path need to take protective action.
Watch
A watch is used when the risk of hazardous weather or hydrologic event has increased significantly, but its concrete location or timing is still uncertain. A watch means that hazardous weather is possible. People should have a plan of action in case a storm threatens and they should be attentive to further information and possible warnings, especially when planning travel or outdoor activities.
Advisory
An advisory is issued when a hazardous weather or hydrologic event is occurring, imminent or likely. Advisories are meant for less serious conditions than warnings, which cause significant inconveniences. However, if caution is not exercised, this could lead to situations that may threaten life or property.
Outlook
An advisory is issued when a hazardous weather or hydrologic event is possible in the next week. Outlooks are intended to raise awareness regarding the potential for significant weather that could lead to situations threatening life or property.

Fig. 12 Warning levels for a typical flood. *Source* National Weather Service, NOAA (2017)

The network was divided into three zones: north (Khyber Pakhtunkhwa and upper Punjab), central (lower Punjab, upper Sindh and Balochistan) and south (lower Sindh) (Fig. 14). The zonation will help in collecting consolidated information from a specific zone and then processing it in the central control location which will, in turn, be used for generation dissemination of information and warnings. A typical sensor node covers a region of 50 km—which should intersect with the radius of the adjacent node in order to receive or transmit information. The topology of the network will

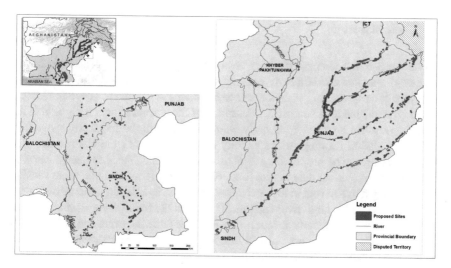

Fig. 13 Proposed geosensor sites (points)

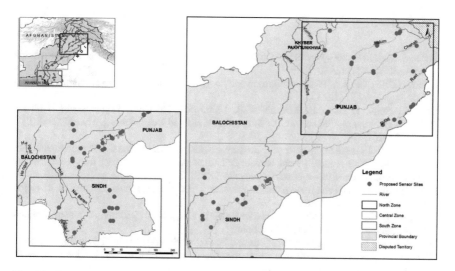

Fig. 14 Geosensor site zonation

be a mesh wherein each node will relay information for another. This information will be transmitted to a central network control which will process the information and generate necessary advice or warning.

There are four levels of warning: the outlook, advisory, watch and warning (Fig. 12). The geosensors network proposed will issues warning based on this mechanism. The information received will be consolidated and processed in the Central Zone, preferably an early flood warning and management organization, based on which the favoured level of warning will be generated and disseminated.

7 Conclusion

As a country that has repeatedly been devastated by floods, Pakistan needs to have a state-of-the-art and efficient early warning system. The present status of early warning systems early warning system needs to be enhanced in order to generate warning based on real-time data analysis. Assessment of the current situation highlights the growing need for a robust flood early warning system in the Indus River majority of the population residing in the vicinity of the river banks are at high risk due to flood hazards. The proposed geosensor network can enhance the early warning capability by providing data in real-time. Flood maps derived from remote sensing imagery play a useful role in the derivation of spatial inundation patterns over time which is suitable for large river catchments such as the Indus River Basin. Geosensor networks generate alerts at four different levels allowing evacuations from high flood risk areas 12–24 hours prior to flooding. They are cost-effective in terms of deployment, maintenance and reliability.

References

Ahmad S, Majeed R (2001) Indus basin irrigation system water budget and associated problems. J Eng Appl Sci 20(1):69–77

Akhtar S (2011) The South Asiatic monsoon and flood hazards in the Indus river basin, Pakistan. J Basic Appl Sci 7(2):101–115

Alfieri L, Burek P, Dutra E, Krzeminski B, Muraro D, Thielen J, Pappenberger F (2013) GloFAS-global ensemble streamflow forecasting and flood early warning. Hydrol Earth Syst Sci 17(3):1161

Ali A (2013) Indus basin floods: mechanisms, impacts, and management: Asian Development Bank

Aon Benfield (2014) Annual global climate and catastrophe report, impact forecasting. http://thoughtleadership.aonbenfield.com/Documents/20140113

Baig MA (2008) Floods and flood plains in Pakistan. In: 20th international congress on irrigation and drainage, Lahore, Pakistan

Basha EEA (2010) In-situ prediction on sensor networks using distributed multiple linear regression models. Massachusetts Institute of Technology

Basha E, Rus D (2007) Design of early warning flood detection systems for developing countries. Paper presented at the Information and communication technologies and development, 2007. International conference on ICTD 2007

Basha EA, Ravela S, Rus D (2008) Model-based monitoring for early warning flood detection. In: SenSys' 08, 5–7 Nov

Benson MA, Matalas NC (1967) Synthetic hydrology based on regional statistical parameters. Water Resour Res 3(4):931–935

Bouwer LM, Bubeck P, Aerts JCJH (2010) Changes in future flood risk due to climate and development in a Dutch polder area. Glob Environ Change 20(3):463–471

Bunruamkaew K, Murayam Y (2011) Site suitability evaluation for ecotourism using GIS & AHP: a case study of Surat Thani province, Thailand. Procedia Soc Behav Sci 21:269–278

Cloke HL, Pappenberger F (2009) Ensemble flood forecasting: a review. J Hydrol 375(3):613–626

Dilts T (2015) New tool—Duration of event for rasters for ArcGIS 10.1. Retrieved from http://gislandscapeecology.blogspot.com/2015/05/new-tool-duration-of-event-forrasters.html on 1 Aug 2016

Dorosh P, Malik SJ, Krausova M (2010) Rehabilitating agriculture and promoting food security after the 2010 Pakistan floods: insights from the south Asian experience. Pak Dev Rev 49:167–192

Federal Flood Commission (2009) Flood management in Pakistan. Annual Flood Report. Retrieved from https://www.mowr.gov.pk/wp-content/uploads/2018/06/Anual-Flood-Report-2009.pdf

Hong Y, Adler RF, Negri A, Huffman GJ (2007) Flood and landslide applications of near real-time satellite rainfall products. Nat Hazards 43(2):285–294

Jiang L, Bergen KM, Brown DG, Zhao T, Tian Q, Qi S (2008) Land-cover change and vulnerability to flooding near Poyang Lake, Jiangxi Province, China. Photogram Eng Remote Sens 74(6):775–786

Jonkman SN, Vrijling JK (2008) Loss of life due to floods. J Flood Risk Manage 1(1):43–56

Lee J-u, Kim J-E, Kim D, Chong PK, Kim J, Jang P (2008) RFMS: real-time flood monitoring system with wireless sensor networks. Paper presented at the 2008 5th IEEE International Conference on Mobile Ad Hoc and Sensor Systems

Libelium (2016) Case studies. Retrieved from Libelium website http://www.libelium.com/resources/case-studies/ on 23 June 2016

Luino F, Cirio CG, Biddoccu M, Agangi A, Giulietto W, Godone F, Nigrelli G (2009) Application of a model to the evaluation of flood damage. Geoinformatica 13(3):339–353

Martinez J-M, Le Toan T (2007) Mapping of flood dynamics and spatial distribution of vegetation in the Amazon floodplain using multitemporal SAR data. Remote Sens Environ 108(3):209–223

Mishra AK, Deep S, Choudhary A (2015) Identification of suitable sites for organic farming using AHP & GIS. Egypt J Remote Sens Space Sci 18(2):181193

Monmonier M (1997) Cartographies of danger: mapping hazards in America. University of Chicago Press, Chicago

Moss ME, Karlinger MR (1974) Surface water network design by regression analysis simulation. Water Resour Res 10(3):427–433

National Oceanographic and Atmospheric Administration (NOAA) (2017) Flood watch information. National weather service. Retrieved from https://www.weather.gov/dvn/floodwatchinfo on April 24, 2017

Nigro J, Slayback D, Policelli F, Brakenridge GB (2014) NASA/DFO MODIS near real time (NRT) global flood mapping product evaluation of flood and permanent water detection. Available online at http://go.nasa.gov/2aS2WuG

Pakistan Meteorological Department (2016) Publications and reports. Retrieved from Pakistan Meteorological Department website http://www.pmd.gov.pk/ on 20 June 2016

Park J, Kwak Y (2011) Determination of inundation area based on flood hazard for a global water risk assessment. Risk Water Resour Manage 347:61–64. IAHS Publications

Pingel N, Watkins D Jr (2009) Multiple flood source expected annual damage computations. J Water Resour Plann Manage 136(3):319–326

Population Association of Pakistan (2016) Demographic indicators of Pakistan. Retrieved from Population Association of Pakistan website http://www.pap.org.pk/statistics/population.htm on 28 July 2016

Pradhan B, Lee S (2009) Delineation of landslide hazard areas on Penang Island, Malaysia, by using frequency ratio, logistic regression, and artificial neural network models. Environ Earth Sci 60(5):1037–1054

Profeti G, Macintosh H (1997) Flood management through Landsat TM and ERS SAR data: a case study. Hydrol Process 11(10):1397–1408

Qi H, Altinakar MS (2011) A GIS-based decision support system for integrated flood management under uncertainty with two dimensional numerical simulations. Environ Model Softw 26(6):817–821

Salma S, Shah MA, Rehman S (2012) Rainfall trends in different climate zones of Pakistan. Pak J Meteorol 9(17):37–47

Shafiq F, Ahsan K (2014) An ICT based early warning system for flood disasters in Pakistan. Res J Recent Sci. ISSN, 2277, 2502

Shaikh IB (2008) Water management for mitigating floods & droughts. In: South-Asian regional workshop on climate change and disaster risk management, Kathmando, Nepal

Simonovic SP (2002) World water dynamics: global modeling of water resources. J Environ Manage 66(3):249–267

Sunkpho J, Ootamakorn C (2011) Real-time flood monitoring and warning system. Sonklanakarin J Sci Technol 33(2):227

Tariq MAUR, Van de Giesen N (2012) Floods and flood management in Pakistan. Phys Chem Earth Parts A/B/C 47:11–20

Ticehurst C, Guerschman JP, Chen Y (2014) The strengths and limitations in using the daily MODIS open water likelihood algorithm for identifying flood events. Remote Sens 6(12):11791–11809

Werner-Allen G, Swieskowski P, Welsh M (2005) Motelab: a wireless sensor network testbed. Paper presented at the Proceedings of the 4th international symposium on Information processing in sensor networks

Zennaro M (2010) Wireless sensor networks for development: potentials and open issues. Doctoral thesis, Stockholm, Sweden

Zhongguo Y, Tianfang C (2015) Wireless sensor network achieved by automatic positioning system node. Int J Future Gener Commun Networking 8(2):93–104

Zlatanova S, Fabbri AG (2009) Geo-ICT for risk and disaster management. In: Geospatial technology and the role of location in science. Springer, Dordrecht, pp 239–266

Chapter 8
Sustaining Local Water Sources: The Need for Sustainable Water Management in the Hill Towns of the Eastern Himalayas

Lakpa Tamang, Ashish Chhetri and Abhijit Chhetri

1 Introduction

Water is a basic necessity for the survival and evolution of human civilization. Although it is ubiquitously distributed, mere availability of water does not guarantee water security in any region—accessibility and quality are sine qua non factors. The access to quality drinking water is considered a powerful determinant of public health (WHO 2006). However, recent studies suggest that shortage of safe drinking water has become a global problem, resulting in an unavoidable global health risk. Recent studies in the Darjeeling Himalayas point towards the perceived impact of climate change, as less snow in the mountains and intense but short episodes of rainfall increase run-off, cause poor accumulation and recharge of water, and ultimately result in the drying up of local water sources (Chaudhary et al. 2011). Studies also indicate a growing perception that climate change-related impacts, which are manifested in the form of an increase in temperatures, more intense precipitation patterns, and longer winter droughts, have further reduced natural groundwater recharge (Tambe et al. 2011). This pattern of shrinking of the monsoon season and resultant drying up of natural springs and declining stream discharge has been recently documented in the Western Himalayas as well (Rawat et al. 2011). The complexities of water ownership in the area also include local dynamics, as major landowners (the military and tea gardens) restrict access to water sources (lakes, river channels, and upland

L. Tamang (✉)
Department of Geography, University of Calcutta, 35, B. C. Road, Kolkata, India
e-mail: ltgeog@caluniv.ac.in

A. Chhetri
Department of Geography, St. Joseph's College, North Point, Darjeeling, India
e-mail: ashishrin30@gmail.com

A. Chhetri
Department of Microbiology, St. Joseph's College, North Point, Darjeeling, India
e-mail: avhijitchhetri@gmail.com

© Springer Nature Switzerland AG 2020
S. Bandyopadhyay et al. (eds.), *Water Management in South Asia*,
Contemporary South Asian Studies, https://doi.org/10.1007/978-3-030-35237-0_8

reserves) and thus exacerbate the need to buy-in water. There is an incompatibility between mainstream federal policy in India and the specificities of life in the hills. For example, the "Million Wells" programme—a flagship antipoverty programme in India—does not benefit the Darjeeling town area wherein freshwater springs, not wells, are the principal source of water (Mell and Sturzakera 2014). Sustainable water harvesting and management of water resources offer the best hope for meeting the challenges of the growing water crisis in the region. For this end, appropriate policy intervention, use of the latest technology, application of new tools such as GIS and information from satellite imageries, community participation, and use of traditional knowledge and water management practices are all essential (Bomjon 2002).

Hence, the management of water resources needs to be taken in the context of an all-regional state-led development and not on a project basis. There must be regional planning in areas such as river basins or watersheds; this in turn should be linked to a long-term national plan (Khawas 2004). Spring sanctuary development could help achieve water availability during the dry season. But investigations must be conducted for a longer term before conclusive results are possible (Negi and Joshi 2002). Moreover, the state government should enable inhabitants to assume greater responsibility for local water management, because this can curb mismanagement and waste. Hence, water management and spring recharge require multi-dimensional approaches, including engineering and biological as well as social and managerial measures.

Water also plays a significant role—potable water supply is needed for domestic, industrial, and public use but water is also utilized for the disposal of sewage waste; all these needs put supplies under tremendous pressure. In the last few decades, pressure has been increasing and greater emphasis is laid on the deterioration in the quality of the spring water. Most springs have been unmindfully used for the disposal of domestic and municipal effluents far beyond their assimilative capacities and have thus been rendered grossly polluted. Despite its importance, water is the most poorly managed resource in the world. The quality of water is becoming vastly deteriorated mainly due to ill-thought out waste disposal; improper water management and care-lessness towards the environment have led to the scarcity of potable water (Agarrkar and Thombre 2009).

Populations living near spring sources use water for domestic household purposes. In addition, there is no frequent and up-to-date monitoring and information providing facilities on the quality of the sewage effluent discharged into either river or spring. Such information is important for the authorities to take proper action so as to prevent pollution and protect public health. Before water can be described as potable, it has to comply with certain physical, chemical, and microbiological standards to ensure that it is palatable and safe for both drinking and other domestic purposes (Subin and Husna 2013).

Darjeeling, a growing hill town, suffers deeply from issues related to water avail-ability, accessibility, and quality (Drew and Rai 2016). The town gets its public water supply from the reservoirs located at the Senchel Wildlife Sanctuary—namely the North Senchel lake with a capacity of 20 million gallons and the South Senchel Lake with a capacity of about 12.5 million gallons. These lakes are fed by natural springs, which constitute the major source of drinking water in the region. Water from the lakes is supplied to the town through public pipelines under the influence of gravity and is stored in several small reservoirs located at different places in the town (Lama and Rai 2016) The water is distributed to different households through a network of numerous inch iron pipes that are controlled by many valves set up at different locations. However, as the water supply system was designed by the British long back—in the 1850s—with much smaller populations to consider (about ten to fifteen thousand during the colonial era), the centralized water supply system is not in a position to meet current water requirements. There is a large gap between demand and supply of water in the Darjeeling town. The daily water demand of the town is 18.6 million gallons, but the supply is only 5.27 million gallons—hence the town suffers a shortage of 13.32 million gallons of water per day (Darjeeling Municipality 2012). It is evident from the above figures that about 60% of the town's population does not have proper access to public water supplies and are compelled to depend on the locally available alternate sources of water (Joshi 2014). Most residents fetch drinking water from the natural springs, generally referred to as *Dhara, Muhan,* and *Simsar,* which are abundantly scattered across the length and breadth of the town. A primary field survey conducted by the authors during 2015–16 traced about 97 natural springs in Darjeeling town along with their GPS points on the real ground surface. However, the locational distribution of the springs is highly uneven, with varying discharging capacities in the town. Among the 32 municipal wards, about 21 wards contain 97% of the spring outlets whereas 11 wards do not have any access to natural springs. The maximum number of natural springs (13) is found in ward No. 6 covering the Merry Villa, Nimki Dara, and Krishna Villa area, while ward Nos. 2, 10,11, 17, 30, and 31 consist of natural springs ranging from 5 to 9 (Fig. 8.1). By contrast, some wards such as Nos. 4, 12, 15, 18, and 19 do not possess a single natural spring.

As a result, there exist huge disparities in terms of water availability and accessibility in different parts of the town. Lack of proper waste and sanitation management in the town have given rise to serious problems relating to spring (groundwater) contamination, particularly faecal contamination due to open disposal of sewerage

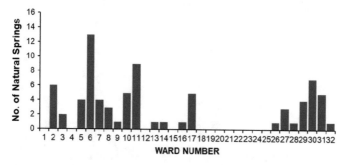

Fig. 8.1 Distribution of natural springs in different wards under the Darjeeling Municipality. *Source* Field survey, 2015–16

to the jhora (drainages) (Drew and Rai 2018). The study aims to investigate the physical (quantitative) and microbial (qualitative) characteristics of spring water in and around Darjeeling town and suggest proper mechanisms for improving the discharging capacity of the springs and thus restore spring water quality. The study relies on the primary investigation of ten major springs selected on the basis of their discharge and dependency and was carried out by the authors during 2015–16. The spring discharge was measured by following simple technique wherein water was collected from a singular point that releases the total availability of spring water in a bottle measuring one litre. The time taken to fill the bottle was recorded from all the selected springs so as to avail a general estimate of water discharge per minute. Similarly, survey in and around the town using questionnaires, focus-group discussions, semi-structured interviews, and informal discussions were also carried out so as to understand the local perception regarding local water sources.

2 Assessment of the Selected Local Water Sources

In order to cover the municipality's overall wards, ten major springs from different locations were selected which also possess high population dependency. The water discharge of different springs was measured in the beginning of every month during 25–30 days. It is evident from the data collected that the discharging capacity of the selected springs is not the same all the year round as it varies greatly from season to season. The discharge of almost all springs depends largely on rainfall patterns and the intensity of surface water infiltration. The environmental surroundings of the springs also influence water availability and the quality of the springs. The springs located in compact settlement areas have low water discharges, whereas the springs located in open areas have comparatively high water discharges.

As shown in Fig. 8.2, water discharge of most springs increases during the monsoon from June to September and decreases steadily during the dry season from October to April. Most springs located on the western slopes have a relatively high water discharging capacity than the springs located on the eastern slope. Most springs

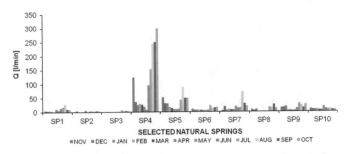

Fig. 8.2 Trend of water discharging capacity of the selected springs. *Source* Field Survey, 2015–16

located on the western slope are characterized by a perennial nature; their water discharge remains almost uniform throughout the year. The springs such as the Giri Dhara (SP4), Lal Dhiki (SP5), and Bhotey Dhara (SP6) have the capacity to provide water throughout the year, with relatively high water discharges ranging from 153.85 L during the monsoon to 35.24 L during the dry season. The population dependency of these perennial springs is very high throughout the year. The possible reasons behind such high discharge of water all year may be the presence of open vegetative cover above the spring outlets. These ridge tops covering the Jalapahar cantonment, the Mahakal Dara, and the Padmaja Naidu Zoological Park may be considered as the major recharge areas for providing surface water infiltration to the perennial springs in the town. The springs of the eastern slopes have low water discharging capacity, and there is a huge range of difference in water discharging capacity between monsoon and lien periods. During dry seasons, many springs have very little discharging capacity—sometimes less than 0.8 L per minute. Some springs such as Bhutia Busty Dhara (SP8) even dry up during the dry season. The reasons behind such situation may be the location of the springs on the escarpment slopes with very little vegetative cover. Such variation in the discharging capacity of springs has led to disparities in terms of water availability and accessibility. Regions having high water discharging springs such as the western slopes may be considered as relatively water-secure areas than the eastern slope which is less secure.

The question of water quality and water health in the Darjeeling municipal area is another significant phenomenon revealing that concerns over water quality are very insignificant in the town. While access to water alone is a challenge in Darjeeling town, the supply of quality drinking water is a rare possibility. The lack of safe drinking water and sanitation services is a serious issue concerning public health and basic amenities in the town. The situation further deteriorates during tourist seasons when the quantity of waste increases manifold (Mell and Sturzakera 2014). The empirical works on the Geo-Socio-Microbiological status as a whole need serious attention so as to combat and minimize water-related diseases in a hill town such as Darjeeling. There still prevail research gaps which have to be filled so as to underlie the extent to which water plays an indispensable role in the sustenance of life—no less as a health determinant, since 80% of diseases in developing countries are due to lack of good quality water, inadequate sanitation, and unavailability of clean drinking water (Cheesbrough 2006).

During our survey to assess the quality of spring water, major parameters such as pH, temperature, faecal coli and total coliforms (MPN/100 ml) have been analysed from water samples collected in pre-sterilized screw-capped borosil bottles every month between November 2015 and October 2016 from the selected springs at the laboratory of the Department of Microbiology, St. Joseph's College, Darjeeling. The results reveal that water in most selected springs is not suitable for drinking purposes. The water temperature in all the selected springs ranges from 14.8 to 19.64 °C. Since temperature plays a vital role in the survival and multiplication of pathogenic bacteria, the highest counts of coliforms are associated with relatively high water temperature. Average pH values in the springs show that the highest pH value is found in the Bhote Dhara (SP7) with about 7.35 while the lowest pH value of 6.3 is found in the Kak

Jhora Dhara (SP3). The microbial analysis shows the presence of minute traces of pathogenic bacteria in almost all the selected springs. The heterotrophic plate count (cfu/ml) of all the samples is found to be greater than 109 cfu/ml on an average, which means the water used for drinking purposes in all the selected springs is not suitable for drinking. The presence-absence (P-A) test for the coliform group was positive in all the selected springs, with little variation. The optional screening intended to isolate the culture of *Faecal Streptococcus and Salmonella* was found to be positive in all the selected springs. The detection of total coliforms and faecal coliforms using LTB, BGLG, and EC broth indicates the presence of coliforms in all the samples and faecal coliforms, which were positive for *E. coli*. The highest number of faecal coliforms of 61 MPN/100 ml is found in the Lal Dhiki Dhara (SP5), followed by the Merry Villa Dhara (SP2) with 56 MPN/100 ml; the lowest is found in Bhutia Busty Dhara (SP8) with 15 MPN/100 ml (Fig. 8.3). Faecal Streptococcus is found to be highest in the Lal Dhiki (SP5) with 118 average count/ml and the lowest is found in the Ghoom Query Dhara (SP1) with 40 average count/ml. Staphylococcus sp. (average count/ml) is found to be highest in the Bhotey Dhara (SP6) with a reading of 116; the lowest of 80 is found in the Dhara Gaun (SP10) (Fig. 8.4).

The overall variation in data distribution shows that different microbial parameters exist in varying degrees at different locations. The table shows that water quality decreases in correlation with decreases in rainfall and temperature. This may be due to the prevalence of maximum downward percolation of pathogenic bacteria into the

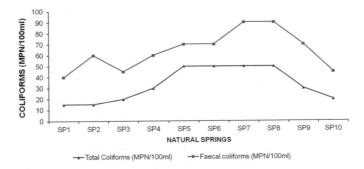

Fig. 8.3 Total and faecal coliforms count of the selected springs. *Source* Field survey, 2015–16

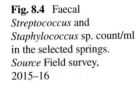

Fig. 8.4 Faecal *Streptococcus* and *Staphylococcus* sp. count/ml in the selected springs. *Source* Field survey, 2015–16

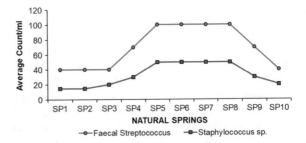

groundwater regulated by the high rate of rainwater infiltration. As mentioned, the recorded temperature of water at all the selected springs (between 14.8 and 19.64 °C) seems favourable for the multiplication of bacteria in the springs. Therefore, it can be observed that the number of bacteria is greater during the warmer months.

The microbiological contamination of water has been traditionally assessed by monitoring concentrations of faecal indicator bacteria such as faecal coliforms and faecal streptococci. The result of all the tests of spring water collected from ten different selected springs proves that the water of the town is highly contaminated with several micro-bacteria—and this is directly correlated with the ever-growing sources of pollutants of various kinds, such as improper waste and sanitation management systems, random disposal of faeces in open drainages, leakages of septic tanks, and open release of untreated wastes from hospitals. In addition to the unsafe biological matter that enters degraded and un-kept pipes, Darjeeling's piped and spring water supplies are subject to cross-contamination from the run-off of 30 to 45 metric tonnes of solid waste that is generated in the Darjeeling Municipality every day (Rai 2011). Other possible sources of contamination include leaking septic systems, surface water leaking into the supply, run-off from agricultural lots, and organically enriched water such as industrial effluents or from decaying plant materials and soils. Such contaminated water can also lead to more serious bacterial infections such as typhoid and diarrhoea. The result of the present bacterial analysis of water shows that the quality of water is not in concordance with the limit prescribed by the WHO and BIS as total coliform counts are above limit. *E. coli* was detected in almost all selected springs during our study; this indicates the recent faecal contamination, which in turn clearly supports the existence of other pathogenic faecal bacteria. The increase in the number of microbial indicators in the spring's water surveyed clearly indicates the possible source of microbial contaminations that can eventually pose health problems. The current practice of water management by local citizens is inadequate with respect to sanitation; it can also be considered hazardous.

3 Major Findings and Discussions

Darjeeling, the "Queen of Hills", is considered as one of the most popular tourist destinations in the state. Hundreds of springs are found in frequent intervals in most of the town and people are heavily dependent on these springs; however, this field survey reveals unawareness regarding the quality of the water available. The water is used simply as a resource, but there have not been concrete steps undertaken so as to examine the quality of water and related health issues. The current scenario of water-related health is not satisfactory due to lack of proper sanitation and drainage, solid waste management system, rapid urbanization, etc. More attention is required to protect and conserve these water sources in and around the town area; this could be achieved by implementing appropriate conservation policies involving regular monitoring and treatment prior to use.

Despite such inevitable necessity, the maintenance of quality of drinking water is highly neglected by the administration. Although studies have been carried out to assess the quality of local potable water; however, it is imperative that comprehensive studies should be undertaken to develop interdisciplinary approaches for the formulation of appropriate conservation and management policies. With the global interest in the potability of drinking water in mind, this study provides a rare opportunity to assess the different aspects of spring water, microbial diversity, ecology, and biochemistry in hill towns.

Acknowledgements The financial support from UGC vide PSW-124/14-15 (ERO) and institutional assistance from St. Joseph's College, Darjeeling are highly acknowledged. The authors would also like to thank Dr. Dorjey Lama and Dr. Upkar Rai for their valuable suggestions during this study.

References

Agarrkar VS, Thombre BS (2009) Status of drinking water quality in schools in Buldhana District of Maharashtra. Nat Environ Pollut Technol 4(1):495–499

Bomjon S (2002) Water resources management in the Eastern Himalayan urban ecosystem, Abstract #2665. EGS XXVII General Assembly, Nice

Chaudhary P, Rai S, Wangdi S, Mao A, Rehman N, Chettri S, Bawa KS (2011) Consistency of local perceptions of climate change in the Kangchenjunga Himalayas landscape. Curr Sci 101(3):504–513

Cheesbrough M (2006) District laboratory Practice in Tropical Countries. Part 2. Cambridge University Press, Cambridge, pp 143–157

Darjeeling Waterworks Department (2012) A report on water supply system of Darjeeling Municipal Area. Darjeeling, India

Drew G, Rai R (2018) Connection amidst disconnection: water struggles, social structures, and geographies of exclusion in Darjeeling. In: Middleton T, Shneiderman S (eds) Darjeeling reconsidered: histories. Oxford University Press, Politics, Environments, pp 219–239

Drew G, Rai R (2016) Water Management in post-colonial Darjeeling: the promise and limits of decentralised resource provision. Asian Stud Rev 40(3):321–339. https://doi.org/10.1080/10357823.2016.1192580

Joshi D (2014) Feminist solidarity? Women, water, and politics in the Darjeeling Himalaya. Mt Res Dev 34(3):243–255. https://doi.org/10.1659/MRD-JOURNAL-D-13-00097.1

Khawas V (2004) Sustainable development and management of water resource in mountain ecosystem: some examples from Sikkim Himalaya

Lama MP, Rai RP (2016) Chokho Pani: an interface between regional and environment in Darjeeling. Himalaya. J Assoc Nepal Himal Stud 36(2):90–98. http://digitalcommons.macalester.edu/himalaya/vol36/iss2/13

Mell C, Sturzakera J (2014) Sustainable urban development in tightly constrained areas: a case study of Darjeeling. Int J Urban Sustain Dev 6(1):65–88. https://doi.org/10.1080/19463138.2014.883994

Negi GCS, Joshi V (2002) Drinking water issues and development of spring sanctuaries in a mountain watershed in the Indian Himalaya. Mt Res Dev 22(1):29–31. https://doi.org/10.1659/0276-4741(2002)022%5b0029:DWIADO%5d2.0.CO;2

Rai RP (2011) Solid waste management in Darjeeling Municipality. In: Desai M, Mitra S (eds) Cloud stone and the mind: the people and environment of Darjeeling hill area. K. P. Bagchi & Company, New Delhi, pp 176–180

Rawat PK, Tiwari PC, Pant CC (2011) Climate change accelerating hydrological hazards and risks in Himalaya: a case study through remote sensing and GIS modelling. Int J Geomat Geosci 1(4):687–699

Subin MP, Husna AH (2013) An assessment on the impact of waste discharge on water quality of Priyar River lets in certain selected sites in the northern part of Ernakulum District in Kerala, India. Int Res J Environ Sci 2(6):76–84

Tambe S, Arrawatia ML, Bhutia NT, Swaroop B (2011) Rapid, cost effective and high-resolution assessment of climate-related vulnerability of rural communities of Sikkim Himalaya, India. Current Sci 101(2):165–173

WHO (2006) Guidelines for safe recreational water environments Volume 2: Swimming pools and similar environments. Geneva, World Health Organization

Part IV
Institutions and Sustainable Regional Use of Water

Chapter 9
Pakistan's Water Economy

Anwar Hussain

1 Introduction

Water is a basic requirement for the development of any country. The majority of
Pakistan's population lives in rural areas and depends—directly or indirectly—on
agricultural sector. The health of the agricultural sector in turn depends on favourable
weather conditions and water availability. In order to meet the growing demand for
food by an increasing population in both rural and urban areas, this sector needs
sufficient water resources and other key inputs. Food production needs to be increased
by 40–50% so as to meet adequate requirements (Associated Press of Pakistan 2016).

Presently, three mountain ranges—the Himalayas, Hindu Kush and Karakoram—
contain 2066 km^3 of ice. This ice is the main source for the Indus River in Pakistan
(Ebrahim 2018), which covers an area of 16.2 million hectares and flows through all
provinces. In Pakistan, around 85% of the annual flows occur in the Kharif (summer)
season and 15% in the Rabi (winter) season (FAO 2018). Pakistan has the world's
largest irrigation system, which consists of three major reservoirs (the Chashma, the
Tarbela and the Mangla), 23 barrages, 12 link canals and 45 irrigation canals (Bhatti
and Kijne 1990; FAO 2018; Government of Pakistan 2013). Pakistan faces severe
water shortages (Altaf et al. 2009; Asif 2013; Faruqui 2004); it has been ranked 3rd
in the list of countries facing water scarcity issues by the International Monetary
Fund (The Express Tribune 2018).

In Pakistan, various factors are responsible for water shortages, including low
efficiency in water use, low storage capacity and mis-management of water resources,
population pressures, climate change, water contamination, nominal water prices,
and lack of effective political governance as well as low investment in the water
sector (Bandaragoda 1998; Bhatti and Farooq 2014; Farooqi et al. 2005; Iqbal 2017b;
Kahlown and Majeed 2003; Qureshi 2011). It is also unfortunate that the national

A. Hussain (✉)
Department of Economics and Development Studies, University of Swat, Swat, Pakistan
e-mail: anwar@uswat.edu.pk

© Springer Nature Switzerland AG 2020
S. Bandyopadhyay et al. (eds.), *Water Management in South Asia*,
Contemporary South Asian Studies, https://doi.org/10.1007/978-3-030-35237-0_9

draft water policy—which could help in achieving the targets and meeting water challenges—has not been approved yet (Kamal 2009). This study focuses on some of these issues.

2 Pakistan's Water Resources and the Threats Involved

Water shortages affect almost all the sectors of Pakistan's economy in general and particularly the agricultural sector, which as mentioned consumes over 93% of the total water resources—a higher number than that present in other South Asian countries (Table 1). Pakistan stands 3rd in the ranking of renewable water resources and 2nd in the ranking of water dependency ratio in South Asia (Table 2).

Water withdrawal per capita is also high when compared to other South Asian countries. Experts hold that its per capita water availability is expected to drop to less than 700 m^3 by 2025 (Murtaza and Zia 2012). Declining per capita availability indicates an increased stress over time (Fig. 1).

The quality and quantity of both surface- and ground-level water resources are declining (Bhutta et al. 2005), and the situation is worsening as time goes by. About 21% of the irrigation requirements in Pakistan depend on groundwater (FAO 2012; Rasul 2014).

Energy is instrumental for increasing agricultural production (Rasul 2014). The current energy crisis can only be tackled through the availability of sufficient water resources. Due to water shortages, cultivable land witnesses desertification (Anjum et al. 2010; Pereira et al. 2002; Rubio et al. 2009). Agricultural exports may be reduced, and the water crisis may create disputes among provinces (Mustafa et al. 2013).

3 Water Footprint and Water Productivity

Currently, Pakistan not only produces water-intensive crops such as rice and sugarcane but also exports these commodities to other countries. Pakistan produces wheat, sugarcane and rice which have the highest water footprint among agricultural crops (Fig. 2). This state of things may push the country into a comprehensive water deficit. Pakistan and India both experienced water deficits during the period 1996–2005. The water footprint linked to grazing and grey water out of total water footprint is in both these countries higher than average in South Asia (Table 3).

There is a need for crops diversification in Pakistan so as to shift from high to low water-consuming crops. An assessment is required so as to suggest appropriate, low water-consuming crops suited for cultivation in the different agroecological zones of Pakistan. The competitiveness of agriculture sector can be improved as a result, which will not only save water but also increase revenue in the agricultural sector.

Table 1 Agricultural area and water withdrawals

Variable	Unit	South Asian Countries						
		Afghanistan	Bangladesh	Bhutan	India	Maldives	Pakistan	Sri Lanka
Water withdrawal by sector								
Agricultural	km³	20.0[1998]	31.5[2008]	0.318[2008]	688[2010]	0.00[2008]	172.4[2008]	11.31[2005]
Municipal	km³	0.2034[2005]	3.6[2008]	0.017[2008]	56[2010]	0.0056[2008]	9.65[2008]	0.805[2005]
Industrial	km³	0.1695[2005]	0.77[2008]	0.003[2008]	17[2010]	0.0003[2008]	1.4[2008]	0.831[2005]
Total	km³	20.28[2000]	35.87[2008]	0.338[2008]	761[2010]	0.0059[2008]	183.5[2008]	12.95[2005]
Total water withdrawal per capita	m³	943.8[2000]	231[2008]	454.5[2008]	602.3[2010]	17.11[2008]	1034[2008]	653.6[2005]
Water withdrawal by source								
Surface water	km³	17.24[1998]	7.39[2008]	0.3379[2008]	396.5[2010]	–	121.9[2008]	–
Groundwater	km³	3.042[1998]	28.48[2008]	0.00[2008]	251[2010]	–	61.6[2008]	–
Total freshwater withdrawal	km³	20.28[2000]	35.87[2008]	0.3378[2008]	647.5[2010]	0.0047[2008]	183.5[2008]	12.95[2005]
Pressure on water resources								
Total freshwater withdrawal as % of TRWR	%	31.04[2000]	2.923[2008]	0.4332[2008]	33.88[2010]	15.67[2008]	74.35[2005]	24.53[2005]
Agricultural water withdrawal as % of TRWR	%	30.61[1998]	2.567[2008]	0.4077[2008]	36[2010]	0.00[2008]	69.85[2008]	21.42[2005]
Total area equipped for irrigation	000 ha	3208[2002]	5050[2008]	31.91[2010]	70,400[2010]	0.00[1998]	19,990[2008]	570[2006]

Source FAO (2018)

Note The figures in brackets represent years

Table 2 Water resources in south Asia (in km³/year, average)

Water resource	South Asian Countries						
	Afghanistan	Bangladesh	Bhutan	India	Maldives	Pakistan	Sri Lanka
Total internal renewable water resources	47.15	105.00	78.00	1446.00	0.03	55.00	52.80
Total external renewable water resources (surface + groundwater)	18.18	1122.00	0.00	464.90	0.00	191.8	0.00
Total renewable water resources	65.33	1227.00	78.00	1911.00	0.03	246.8	52.80
Dependency ratio (%)	28.72	91.44	0.00	30.52	0.00	77.71	0.00

Source FAO (2017)

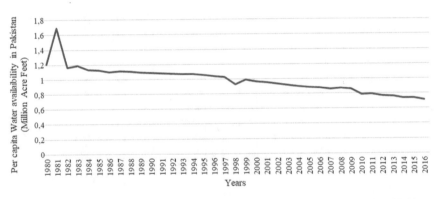

Fig. 1 Trends of water scarcity in Pakistan (1985–2016). *Source* Agricultural Statistics of Pakistan

As a matter of fact, Pakistani agriculture is mainly driven by the livestock sector, which is growing at 3.8% per year and contributes 59% to the agricultural GDP (11% to Pakistan's overall GDP). The total population of cattle, buffalos, sheep and goats rounds 46.1, 38.8, 30.5 and 74.1 millions respectively (Government of Pakistan 2018). The country not only possesses such large numbers of livestock but also exports livestock products to other countries—which is also a key factor of virtual water trade deficit in Pakistan. This is because livestock products such as beef, nuts, sheep/goat meat, pig meat and butter have the higher water footprint among all agricultural products (Fig. 3).

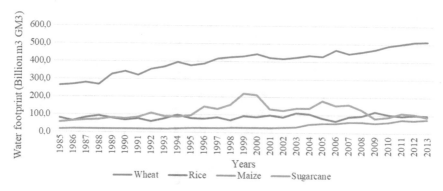

Fig. 2 Water footprint of selected crops in Pakistan 1985–2013

There is also a need to improve both the physical[1] and economic water productivity in Pakistan; this must be accomplished through better water control and improved production processes. To this end, one option is to specialize in those crops which require less water and have guaranteed markets both nationally or globally. However, in Pakistan sugarcane and rice constitute high water-consuming (Fig. 4), established cash crops. Besides substitution, there is also need to cultivate low water-consuming crops in unirrigated areas or cultivable wastelands. In Pakistan, an area of about 8 million hectares constitutes cultivable wasteland which remains uncultivated due to various factors—including low water availability. The current cropping pattern should be adjusted to fit water availability characteristics across different agroecological zones of Pakistan. The Pakistan Agricultural Research Council (PARC) and the Pakistan Council of Research in Water Resources (PCRWR) can move so as to identify suitable crops for different agroecological zones while also considering the water requirements of each crop in each zone.

4 Water Capacity, Management and Governance

Pakistan has a limited water storage capacity (Ashraf et al. 2007; Kahlown and Majeed 2003; Keller et al. 2000) and the heavy rainfall water monsoon destroys both agricultural land and hundreds of livelihoods and lives almost every year (Adnan 2009; Akhtar 2011). Floods constituted 46.8% of all disasters occurring in Pakistan and were responsible for 1.5% of recorded mortality. They also caused 73% of the country's economic losses—amounting to $1029 million during the period 1990–2014 (Shariff 2016).

As a matter of fact, due to such low water storage capacity, flood water cannot presently be saved for productive purposes. Water with an estimated economic value

---

[1]Physical water productivity is a term signifying the production of more output per unit of water; the term economic water productivity represents the revenue obtained from crops per unit of water.

Table 3 Water footprint of national production (mm³/yr) 1996–2005

	Green	Blue	Grey	Green	Blue	Blue	Grey	Blue	Grey	Green	Blue	Grey	Green	Blue	Grey
Afghanistan	7856	7664	149	4043	132	0.1	1	42	378	11,899	7838	528	1114.6	8.7	236.4
Bangladesh	62,919	7825	10,138	3342	434	16.5	313.5	401.6	3614.4	66,261	8677	14,066	5859.5	3192.2	469.1
Bhutan	624	27	2	74	8	0.3	4.8	2	18	698	37	25	41.8	23.4	15.4
India	716,004	231,428	99,429	42,644	4707	1760.5	33,449.5	5224	47,016	758,649	243,119	179,895	−60,505	−18,454	−16,449
Maldives	66	2	3	0	0	0	0	0.3	1.9	66	2	5	172.1	26	36.9
Pakistan	40,561	74,272	21,805	34,113	907	173.5	3296.5	327	2943	74,674	75,679	28,044	−863.2	−31,369	−9504.1

Source Mekonnen and Hoekstra (2011)

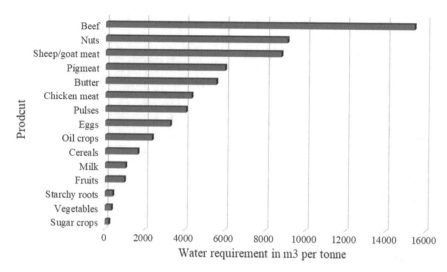

Fig. 3 Water requirements for different products in m^3 per tonne. *Source* Mekonnen and Hoekstra (2012)

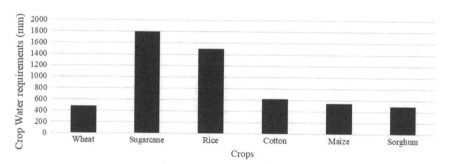

Fig. 4 Crop water requirements in Pakistan. *Source* Pakistan Council of Water Resource (PCWR) research report 2 (2003)

of approximately seventy billion dollars is being thrown unused in the sea every year (Iqbal 2017a). Pakistan can barely store thirty days of water in the major reservoir of the Indus Basin Irrigation System (Hussain 2017). In Pakistan, the Tarbela and Mangla dams have the highest reservoir capacity (Table 4). Low water capacity has various implications—among them the fact that the quality of the water may deteriorate and spread many waterborne diseases (Thacker et al. 1980).

As most of the rural population (65.9%) in Pakistan is engaged in agricultural activities (Bhutto and Bazmi 2007), non-availability of water would certainly affect employment levels (Nizamani et al. 1998). Decline in water availability can further decrease agricultural production and lead to a hike in food prices.

Agricultural production also requires energy for irrigating and processing food commodities. In Pakistan, the key source of energy is hydrological—low water

Table 4 Water reservoir capacity of Pakistan

Name of dam	Administrative unit	River	Major basin	Completed/operational since	Dam height (m)	Reservoir capacity (million m³)	Reservoir area (km²)	Decimal degree latitude	Decimal degree longitude
Namal		Golar	Indus	1913	32.00	27.60	2.20	32.69	71.80
Sukkur Barrage	Sindh	Indus	Indus	1932				27.68	68.85
Trimmu Barrage		Chenab		1939				31.14	72.15
Taunsa Barrage	Punjab	Indus	Indus	1958				30.51	70.86
Kalri						2282.00	100.00	24.84	67.99
Kotri Barrage	Sindh							25.44	68.32
Panjnad Barrage	Punjab							29.35	71.02
Sulemanki		Sutlej	Indus						
Warsak	NWFP	Kabul	Indus	1960	72.00	76.50	8.60	34.16	71.36
Jinnah		Indus	Indus			231.56	8.80	32.92	71.51
Islam		Sutlej	Indus			141.11	5.20	29.83	72.55
Balloki		Ravi	Indus					31.22	73.86
Ferozepur		Sutlej	Indus			21.38	0.70	30.99	74.55
Guddu Barrage	Sindh			1962				28.41	69.72
Baran		Baran	Indus	1962	24.00	120.88	6.70	33.01	70.50

(continued)

Table 4 (continued)

Name of dam	Administrative unit	River	Major basin	Completed/operational since	Dam height (m)	Reservoir capacity (million m³)	Reservoir area (km²)	Decimal degree latitude	Decimal degree longitude
Rawal	Capital Territory			1962	40.58	54.15		33.69	73.12
Misriot	Punjab			1963	14.21	0.69			
Sipiala	Punjab			1964	11.42	0.70			
Sidhnai Barrage		Ravi	Indus	1965				30.57	72.16
Tanda	NWFP	Kohat Toi	Indus	1967	35.00	96.71	1.80	33.57	71.40
Dhurnal	Punjab			1967	20.67	1.70		32.81	72.10
Rasul		Jhelum	Indus	1967	26.00	132.25	20.90	32.68	73.52
Mangla	AJK	Jhelum (a tributary of Indus)	Indus	1967	116.00	10,147.59		33.15	73.64
Qadirabad	Punjab	Chenab	Indus	1968	22.00	178.04	10.20	32.32	73.69
Marala	Punjab	Chenab	Indus	1968	5.00			32.67	74.47
Nirali	Punjab			1970	20.82	0.84			
Ratti Kassi	Punjab			1970	14.27	2.40			
Gurabh	Punjab			1970	21.19	0.84		32.86	72.36
Chashma	Punjab	Indus	Indus	1971		870.00		32.44	71.38
Qibla Bandi	Punjab			1971	21.28	2.24			
Dungi	Punjab			1971	21.70	2.17		33.29	73.30
Kanjoor	Punjab			1975	18.62	3.50			

(continued)

Table 4 (continued)

Name of dam	Administrative unit	River	Major basin	Completed/operational since	Dam height (m)	Reservoir capacity (million m³)	Reservoir area (km²)	Decimal degree latitude	Decimal degree longitude
Tarbela	NWFP	Indus	Indus	1976	137.00	11,960.10	260.00	34.09	72.70
Hub	Sindh	Hub	Indus	1979	49.00	1139.80	13.90	25.24	67.11
Chhanni Bar	Punjab			1979	19.45	2.41			
Khokhar zer	Punjab			1979	23.40	4.08			
Garat	Punjab			1981	20.06	2.75		33.02	73.52
Walana	Punjab			1983	21.28	2.70			
Khanpur	Punjab	Haro River		1983	50.75	130.70		33.80	72.93
Simly	Capital Territory			1983		34.28		33.72	73.34
Khasala	Punjab			1985	18.24	2.98			
Surla	Punjab			1985	18.54	2.35			
Shahpur	Punjab			1986	25.84	17.66			
Mirwal	Punjab			1990	24.01	4.64			
Bhugtal	Punjab			1990	22.80	1.41			
Nikka	Punjab			1990	29.48	1.54			
Dhok Sanday Mar	Capital Territory			1990	14.59	0.80			
Jabbi	Punjab			1991	10.76	3.33			
Dhok Outab Din	Punjab			1991	24.62	2.14			
Kot Raja	Punjab			1991	24.05	3.51			

(continued)

Table 4 (continued)

Name of dam	Administrative unit	River	Major basin	Completed/operational since	Dam height (m)	Reservoir capacity (million m^3)	Reservoir area (km^2)	Decimal degree latitude	Decimal degree longitude
Jammargal	Punjab			1992	16.26	3.00			
Jawa	Punjab			1994	25.05	1.94			
Shakar Dara	Punjab			1994	34.95	7.00			
Tain Pura I	Punjab			1994	25.28	7.33			
Tain Pura II	Punjab			1994	24.27	4.19			
Pira Fatehal	Punjab			1995	27.34	9.12			
New Dhok Tahlian	Punjab			2002	25.36	2.23			
Basal	Punjab			2004	18.66	2.08			
Mial	Punjab			2004	21.37	3.95			
Jabba	Punjab			2005	25.41	1.06			
Jalwal	Punjab			2005	18.24	6.17			
Sawal	Punjab			2005	28.88	2.96			
Talikna	Punjab			2005	17.59	2.53			
Thattl Syedan	Punjab			2005	12.96	0.74			
Jamal	Punjab			2005	26.44	2.29			
Lehri	Punjab			2005	33.13	7.03			
Salial	Punjab			2005	20.67	0.65			
Ghazial	Punjab			2007	22.34	2.47			

(continued)

Table 4 (continued)

Name of dam	Administrative unit	River	Major basin	Completed/operational since	Dam height (m)	Reservoir capacity (million m^3)	Reservoir area (km^2)	Decimal degree latitude	Decimal degree longitude
Khai	Punjab			2007	38.91	7.30			
Phalina	Punjab			2008	22.49	4.81			
Minwal	Punjab			2008	25.08	2.47			
Domeli	Punjab			2008	36.47	10.71			
Fatehpur	Punjab			2008	26.29	2.14			
Shah Habib	Punjab			2008	23.45	2.04			

Source FAO (2018)

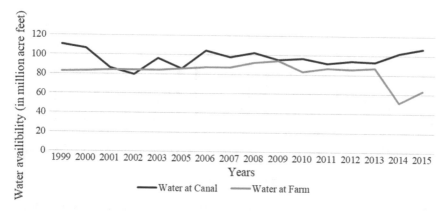

Fig. 5 Gap between water available in a canal and that available to a farm in Pakistan. *Source* Agricultural Statistics of Pakistan (2015)

storage capacity thus affects national energetic efficiency levels and leads to low agricultural productivity (Siddiqi 2018).

The construction of mega dams can solve low water storage capacity in the country with relative ease (Ashraf et al. 2007; Jamali et al. 2014; Keller et al. 2000). However, existing water can also be better administered. In Pakistan, there is a gap between the water available in canals and at farm gate level (Fig. 5). Water losses due to leakages can be controlled through water cemented channels and by strictly regulating water distribution channels. A decentralized water governance system can best solve such issues.

The current water crisis in Pakistan also demands the development of groundwater potential (Qureshi 2011). In some areas, groundwater resources are over-exploited (Commission on Science and Technology for Sustainable Development in the South 2003). Excessive groundwater extraction from tube wells through solar water pumping in rural areas needs to be controlled. Farmers extract groundwater using cheap electricity. The key issue with groundwater depletion is also the lack of proper rights as regards consuming and producing groundwater resources (Holden and Thobani 1999; Rosegrant and Binswanger 1994). Currently, there are no regulations regarding the size, length and space for the installation of tube wells. Alternatively, the *karez system* in Baluchistan can be rehabilitated for a sustainable water supply (Memon et al. 2017).

Imbalanced water distribution patterns are also a key cause of water waste in the country. At field level, the influential takes more water than the poor. The water is allocated mainly on the basis of the watercourse's cultivable command area (Bhatti and Kijne 1990). The prevailing "Warabandi"[2] system sometimes also causes water waste because farmers do not need water for their agricultural lands every single

---

[2]In the Warabandi system, each farmer has a turn being provided with fixed water time. A farmer' turn may come once a week.

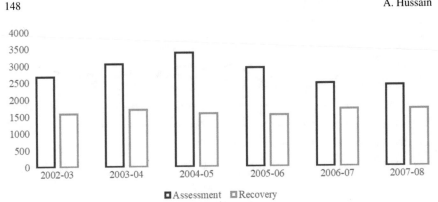

Fig. 6 Assessment and collection of Abyana in Pakistan (Rs. Million). *Source* Government of Pakistan (2012)

time it is made available, and the same water is ultimately lost. So water distribution may be made flexible so as to improve efficiency in water use.

Pakistan also has low water use efficiency (Howell 2001; Kahlown and Majeed 2003; Mohammad et al. 2012) which capacity is 40% (Bakhsh et al. 2015). This can be improved through effective water governance, regulations and quality inputs.

5 Water Pricing

Water markets are meant to allocate scarce water resources efficiently. There is an urgent need to set water prices on the basis of demand and supply dynamics (Altaf et al.1993). In Pakistan, available water prices are determined on the flat rate system, i.e. in Pak rupees per unit of area. These water charges (locally termed as Abyana) remain fixed for many years. There also remain gaps in the Abyana assessment and their actual collection practices (Fig. 6). The Abyana collected in the year 2012 was 60% of the total assessed. Abyana collection recovers only 24% of the total operating and maintaining costs; this calls for the reinforcement of financial sustainability in Pakistan (Government of Pakistan 2012).

6 Water and Ecosystem Services

The services provided by various ecosystems are bound to the availability of sufficient quantity and quality water resources (Brouwer et al. 2013; Dangles et al. 2004). For example, wetlands provide many services (Clarkson et al. 2013) such as protecting biodiversity, storing and filtering water resources and providing medicinal plants and recreation in many areas of Pakistan. However, due to non-availability of water resources, these wetlands are dying. The tourist spots which were the key source of income for millions—and also a significant source of national income—are being affected by a shortage of water. The forests, fisheries and water bodies do not receive water in sufficient quality and quantity to function properly (Grizzetti et al. 2016; Wantzen et al. 2008).

7 Water and Health

The provision of both quality and quantity—appropriate water—is very important. However in Pakistan, about 25–30 of total patients admitted in the healthcare system present waterborne diseases—and no less than 60% of all infants die due to infected water (Butt and Khair 2016). Water, sanitation and hygiene-related diseases cost Pakistan 112 billion rupees (Hakro 2012). Health costs at household level in both rural and urban areas will further increase if no effective measures are taken.

The present domestic water use in Pakistan is 8%, and most urban areas depend on groundwater for water supply. In rural areas, only 53% of the population has access to safe drinking water; the remaining population gets its water from untreated water supplies. According to the WHO, 80% of diseases are caused by unsafe drinking water and unhygienic conditions. The provision of proper sanitation and healthy drinking water are indispensable. In Pakistan, over 65% population has access to safe drinking water (Khan 2012).

In Pakistan, about half the population has access to at least basic sanitation facilities (Table 5)—the number in Bhutan is 62%, in the Maldives 95% and in Sri Lanka 94% (Joint Monitoring Programme 2018). About half of Pakistan's population has access to drinking water at a basic service level (Table 5); in Bhutan, the number is 63%. So, there is an urgent need to provide sanitation facilities as well as drinking water and hygienic infrastructure to both rural and urban populations.

Increasing water pollution and wastewater is another factor causing many health problems in Pakistan (Azizullah et al. 2011). The rural residential sector contributes 48% of the total wastewater generated (Fig. 7). In 2000, the total wastewater produced was estimated 12.33 km^3, while treated wastewater was an estimated 0.145 km^3 (FAO 2018). There is a dire need to properly treat this contaminated water. Polluted and urban wastewater can also be recycled for irrigation purposes; however, this requires heavy investments.

8 Water and Institutional Arrangements

Proper water institutions are instrumental in the goal for better water management. There are many national and international institutions working in Pakistan in water-related issues. The details regarding the type, sectors and activities of these institutions are reported in Table 6. However, there is still a need to strengthen these

Table 5 Status of sanitation, drinking water and hygiene in Pakistan (2015)

Area level	Service type	Coverage (%)	Service level
Total	Sanitation	58.25	At least basic
Rural	Sanitation	48.05	At least basic
Urban	Sanitation	74.37	At least basic
Total	Drinking water	52.91	Basic service
Rural	Drinking water	54.22	Basic service
Urban	Drinking water	50.83	Basic service
Total	Hygiene	60.47	Basic service
Rural	Hygiene	46.05	Basic service
Urban	Hygiene	83.25	Basic service
Total	Hygiene	31.24	Limited service
Rural	Hygiene	43.34	Limited service
Urban	Hygiene	12.12	Limited service
Total	Sanitation	8.45	Limited service
Rural	Sanitation	8.90	Limited service
Urban	Sanitation	7.73	Limited service
Total	Drinking water	2.72	Limited service
Rural	Drinking water	3.41	Limited service
Urban	Drinking water	1.64	Limited service
Total	Hygiene	8.29	No hand washing facility
Rural	Hygiene	10.61	No hand washing facility
Urban	Hygiene	4.64	No hand washing facility
Total	Sanitation	11.55	Open defecation
Rural	Sanitation	18.85	Open defecation
Urban	Sanitation	0.00	Open defecation
Total	Drinking water	35.64	Safely managed service
Rural	Drinking water	32.42	Safely managed service
Urban	Drinking water	40.72	Safely managed service
Total	Drinking water	2.45	Surface water
Rural	Drinking water	3.86	Surface water
Urban	Drinking water	0.23	Surface water
Total	Sanitation	21.76	Unimproved
Rural	Sanitation	24.20	Unimproved
Urban	Sanitation	17.90	Unimproved
Total	Drinking water	6.28	Unimproved
Rural	Drinking water	6.09	Unimproved
Urban	Drinking water	6.58	Unimproved

Source Joint Monitoring Programme (2018)

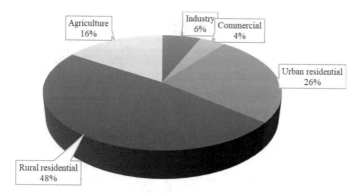

Fig. 7 Wastewater status in Pakistan. *Source* Murtaza and Zia (2012)

institutions' capacities. The private sector also needs to be encouraged to cooperate with the government so as to tackle the current water crisis.

The devolution of water institutions and policies across provinces in Pakistan should not be ambiguous and each institution, whether federal or provincial, should freely and actively work in their bounded circles—with clear responsibilities without any political pressures. Some of the water storage issues such as the construction of the Kalabagh Dam have become more of a political than economic dynamic.

9 Transboundary Water Issues

The well-known Indus Water Treaty was signed in 1960 between Pakistan and India. Accordingly, the control of the flow of three rivers—namely the Bias, Ravi and Sutlej—was apportioned to India while the control of Indus, Chenab and Jhelum was apportioned to Pakistan. India is constructing two power plants, the Kishenganga and Ralte; this action has been challenged by Pakistan (Ashfar 2017). The construction by India of dams affecting the flow of water in Pakistan has serious implications for the country's agricultural sector (Qureshi 2011). There is a need to solve this dispute on water distribution so as to ensure sufficient access to water (Sahni 2006). The Indus River system also receives about 15% of its water through the River Kabul flowing in Pakistan (Lead 2017). Both Afghanistan and Pakistan would need a general agreement over such water issues as regards both present and future.

Table 6 Water institutional arrangements in Pakistan

Name	Type	Sectors	Activities
Water and Power Development Authority (WAPDA) child of: Ministry of Water and Power	University/research institution	Dams, Irrigation, Municipalities, Natural resources	Infrastructure development
Water and Power Development Authority (WAPDA)	Government institution	Dams	Infrastructure development, operation and maintenance
Irrigation Department, Punjab child of: Government of Punjab, Lahore	Government institution	Irrigation	Infrastructure development, operation and maintenance
Irrigation Department, Sindh child of: Government of Sindh, Karachi	Government institution	Irrigation	Infrastructure development, operation and maintenance
Irrigation Department, Government of Khyber Pakhtunkhwa	Government institution	Irrigation	Infrastructure development, operation and maintenance
Irrigation and Power Department, Balochistan child of: Government of Balochistan, Quetta	Government institution	Dams, Irrigation	Infrastructure development, operation and maintenance
Indus River System Authority (IRSA) child of: Ministry of Water and Power	Government institution	Transboundary water, Water	Licensing and allocation, Policy and strategy
Federal Water Management Cell (FWMC) child of: Ministry of National Food Security and Research	Government institution	Irrigation, Municipalities, Natural resources, Water	Licensing and allocation, Policy and strategy
Ministry of Water and Power (MoWP)	Government institution	Dams, Water	Policy and strategy

(continued)

Table 6 (continued)

Name	Type	Sectors	Activities
Centre of Excellence in Water Resources Engineering and Management (CEWREM) child of: University of Engineering and Technology, Lahore	University/research institution	Irrigation, Municipalities, Natural resources	Policy and strategy, Research
Ministry of National Food Security and Research	Government institution	Agriculture, Water	Policy and strategy, research, statistics and monitoring
Pakistan Agricultural Research Council (PARC) child of: Minsitry of National Food Security and Research	University/research institution	Agriculture	Research
Pakistan Council of Research in Water Resources (PCRWR) child of: Ministry of Science and Technology	University/research institution	Irrigation, Municipalities, Natural resources	Research
National Agricultural Research Centre (NARC) child of: Pakistan Agricultural Research Council	University/research institution	Agriculture	Research
Water Resources Research Institute (WRRI) child of: National Agricultural Research Centre (replaced by: Climate Change, Alternate Energy and Water Resources Institute)	University/research institution	Irrigation, Municipalities, Natural resources	Research

(continued)

Table 6 (continued)

Name	Type	Sectors	Activities
International Waterlogging and Salinity Research Institute, Lahore, Pakistan (IWASRI) child of: Water and Power Development Authority (WAPDA)	University/research Institution	Irrigation, Natural resources	Research
Climate Change, Alternate Energy and Water Resources Institute (CAEWRI) child of: Pakistan Agricultural Research Council	University/research institution	Agriculture	Research
Soil Survey of Pakistan (SSP)	Government institution	Agriculture, Natural resources	Statistics and monitoring
Pakistan Meteorological Department (PMD) child of: Ministry of Defence, Government of Pakistan	University/research institution	Environment, Water	Statistics and monitoring

Source FAO (2016) online database

10 The Way Forward

From the above discussion, it is concluded that Pakistan faces serious water challenges. Per capita water availability is declining. Most water is used by the agricultural sector which depends on producing and exporting high water-consuming products. The capacity of water storage is low. Energy insecurity in the country is mainly driven by issues of water insecurity. Imbalanced water distribution, pricing, leakages, groundwater depletion, wastewater recycling, water use efficiency, worsening ecosystem services and low quality of drinking water, sanitation and hygiene facilities and related health costs are all key water-related challenges in Pakistan. There is an urgent need to cope with these challenges by constructing mega reservoirs. The agricultural sector should also rely on low water-consuming crops. The governance of the water sector needs improvement so as to tackle the issues of imbalanced water distribution, pricing and leakages. Water use efficiency needs to be improved. Wastewater can be recycled for agricultural purposes. Water provision for an adequate functioning of ecosystem services should be ensured. Provision of basic health facilities covering drinking water, sanitation and hygiene should be prioritized.

References

Adnan S (2009) Effective rainfall for irrigated agriculture plains of Pakistan. Pak J Meteorol 6(11):61–72

Akhtar S (2011) The South Asiatic monsoon and flood hazards in the Indus river basin, Pakistan. J Basic Appl Sci 7(2):101–115

Altaf MA, Whittington D, Jamal H, Smith VK (1993) Rethinking rural water supply policy in the Punjab, Pakistan. Water Resour Res 29(7):1943–1954

Altaf S, Kugelman M, Hathaway RM (2009) Running on empty. Pakistan's water crisis: Woodrow Wilson International Center for Scholars

Anjum SA, Wang L-c, Xue L, Saleem MF, Wang G-x, Zou C-m (2010) Desertification in Pakistan: causes, impacts and management. J Food Agric Environ 8(2):1203–1208

Ashfar M (2017) India-Pakistan water dispute. The Nation. Retrieved from https://nation.com.pk/22-Sep-2017/india-pakistan-water-dispute

Ashraf M, Kahlown MA, Ashfaq A (2007) Impact of small dams on agriculture and groundwater development: a case study from Pakistan. Agric Water Manag 92(1–2):90–98

Asif M (2013) Climatic change, irrigation water crisis and food security in Pakistan

Associated Press of Pakistan (2016) Pakistan must increase food production by 40–50%, The express tribune. Retrieved from https://tribune.com.pk/story/1063592/pakistan-must-increase-food-production-by-40-50/

Azizullah A, Khattak MNK, Richter P, Häder D-P (2011) Water pollution in Pakistan and its impact on public health—a review. Environ Int 37(2):479–497

Bakhsh A, Ashfaq M, Ali A, Hussain M, Rasool G, Haider Z, Faraz R (2015) Economic evaluation of different irrigation systems for wheat production in Rechna Doab, Pakistan. Pak J Agric Sci 52(3):821–828

Bandaragoda DJ (1998) Design and practice of water allocation rules: lessons from Warabandi in Pakistan's Punjab, vol 17. IWMI

Bhatti M, Farooq M (2014) Politics of Water in Pakistan. Pak J Soc Sci (PJSS) 34(1):205–216

Bhatti MA, Kijne J (1990) Irrigation allocation problems at tertiary level in Pakistan. Working Paper No. 90.3c. Overseas Development Institute and International Irrigation Management Institute

Bhutta M, Chaudhry M, Chaudhry A (2005) Groundwater quality and availability in Pakistan. Paper presented at the Proceedings of the seminar on strategies to address the present and future water quality issues

Bhutto AW, Bazmi AA (2007) Sustainable agriculture and eradication of rural poverty in Pakistan. Paper presented at the Natural Resources Forum

Brouwer R, Hassan R, Beukering P, Papyrakis E, Bouma J (2013) Water—related ecosystem services. In: Nature's wealth: the economics of ecosystem services and poverty, p 259

Butt M, Khair SM (2016) Cost of illness of water-borne diseases: a case study of Quetta. J Appl Emerg Sci 5(2):133–143

Clarkson BR, Ausseil A-GE, Gerbeaux P (2013) Wetland ecosystem services. In: Ecosystem services in New Zealand: conditions and trends. Manaaki Whenua Press, Lincoln, pp 192–202

Commission on Science and Technology for Sustainable Development in the South (2003) Water-resources situation in Pakistan: challenges and future strategies. COMSATS' Series of Publications on Science and Technology

Dangles O, Gessner MO, Guérold F, Chauvet E (2004) Impacts of stream acidification on litter breakdown: implications for assessing ecosystem functioning. J Appl Ecol 41(2):365–378

Ebrahim ZT (2018) Is Pakistan running out of fresh water? Retrieved 28 June 2018

FAO (2012) Irrigation in Southern and Eastern Asia in Figures AQUASTAT Survey—2011. FAO Water reports 37. Food and Agriculture Organization of the United Nations, Rome

FAO (2018) Water resources. http://www.fao.org/nr/water/aquastat/countries_regions/PAK/

Farooqi AB, Khan AH, Mir H (2005) Climate change perspective in Pakistan. Pak J Meteorol 2(3)

Faruqui NI (2004) Responding to the water crisis in Pakistan. Int J Water Resour Dev 20(2):177–192

Government of Pakistan (2012) Canal water pricing in Pakistan: Assessment, issues and options. Islamabad

Government of Pakistan (2013) Review of National Water Policy Documents and Strategies in the Context of SLM, NAP and UNCCD. Ministry of Climate Change Islamabad

Government of Pakistan (2018) Pakistan economic survey (2017–18)

Grizzetti B, Lanzanova D, Liquete C, Reynaud A, Cardoso A (2016) Assessing water ecosystem services for water resource management. Environ Sci Policy 61:194–203

Hakro AN (2012) Water, sanitation and poverty linkages in Pakistan. Asian J Water Environ Pollut 9(3):25–36

Holden P, Thobani M (1999) Tradable water rights: a property rights approach to resolving water shortages and promoting investment: The World Bank

Howell TA (2001) Enhancing water use efficiency in irrigated agriculture. Agron J 93(2):281–289

Hussain A (2017) Water under threat. The NEWS. Retrieved from https://www.thenews.com.pk/print/93102-Water-under-threat

Iqbal Z (2017a) Water scarcity in Pakistan. The Nation. Retrieved from https://nation.com.pk/07-Dec-2017/water-scarcity-in-pakistan

Iqbal Z (2017b) Water scarcity in Pakistan—causes, effects and solutions. The Nation. Retrieved from https://nation.com.pk/11-Dec-2017/water-scarcity-in-pakistan-causes-effects-and-solutions

Jamali IA, Mörtberg U, Olofsson B, Shafique M (2014) A spatial multi-criteria analysis approach for locating suitable sites for construction of subsurface dams in Northern Pakistan. Water Resour Manage 28(14):5157–5174

Joint Monitoring Programme (2018) WHO/UNICEF joint monitoring programme for water supply, sanitation and hygiene (JMP) https://washdata.org/

Kahlown M, Majeed A (2003) Water-resources situation in Pakistan: challenges and future strategies. Water resources in the South: present scenario and future prospects, p 20

Kamal S (2009) Pakistan's water challenges: entitlement, access, efficiency, and equity. Running on Empty

Keller AA, Sakthivadivel R, Seckler DW (2000) Water scarcity and the role of storage in development, vol 39. IWMI

Khan A (2012) Natural resources of Pakistan. Economics and Education. Retrieved from: http://ahsankhaneco.blogspot.com/2012/04/natural-resources-of-pakistan.html on 20 July 2019

Lead (2017) Indus water treaty & Pakistan's water diplomacy. Retrieved 1 July 2018, from http://www.lead.org.pk/lead/postDetail.aspx?postid=341

Mekonnen MM, Hoekstra AY (2011) National water footprint accounts: the green, blue and grey water footprint of production and consumption. Value of Water Research Report Series No. 50, UNESCO-IHE, Delft, the Netherlands

Mekonnen MM, Hoekstra AY (2012) A global assessment of the water footprint of farm animal products. Ecosystems 15(3):401–415

Memon JA, Jogezai G, Hussain A, Alizai MQ, Baloch MA (2017) Rehabilitating traditional irrigation systems: assessing popular support for Karez rehabilitation in Balochistan, Pakistan. Hum Ecol 45(2):265–275

Mohammad W, Shah S, Shehzadi S, Shah S (2012) Effect of tillage, rotation and crop residues on wheat crop productivity, fertilizer nitrogen and water use efficiency and soil organic carbon status in dry area (rainfed) of north-west Pakistan. J Soil Sci Plant Nutr 12(4):715–727

Murtaza G, Zia MH (2012) Wastewater production, treatment and use in Pakistan. Retrieved 25 June 2018, from http://www.ais.unwater.org/ais/pluginfile.php/232/mod_page/content/127/pakistan_murtaza_finalcountryreport2012.pdf

Mustafa D, Akhter M, Nasrallah N (2013) Understanding Pakistan's water-security nexus. United States Institute of Peace Washington, DC

Nizamani A, Rauf F, Khoso AH (1998) Case study: Pakistan. Population and water resources

Pereira LS, Oweis T, Zairi A (2002) Irrigation management under water scarcity. Agric Water Manag 57(3):175–206

Qureshi AS (2011) Water management in the Indus basin in Pakistan: challenges and opportunities. Mt Res Dev 31(3):252–260

Rasul G (2014) Food, water, and energy security in South Asia: a nexus perspective from the Hindu Kush Himalayan region☆. Environ Sci Policy 39:35–48

Rosegrant MW, Binswanger HP (1994) Markets in tradable water rights: potential for efficiency gains in developing country water resource allocation. World Dev 22(11):1613–1625

Rubio J, Safriel U, Daussa R, Blum W, Pedrazzini F (2009) Water scarcity, land degradation and desertification in the Mediterranean Region: environmental and security aspects. Springer Science & Business Media

Sahni HK (2006) The politics of water in South Asia: the case of the indus waters treaty. SAIS Rev Int Aff 26(2):153–165

Shariff K (2016) Preparing for the monsoons and beyond. The Nation. Retrieved from https://nation.com.pk/01-Jun-2016/preparing-for-the-monsoons-and-beyond

Siddiqi A (2018) The water-energy-food nexus of Pakistan. Retrieved 29 June 2018, from http://www.pakissan.com/english/issues/water.energy.food.nexus.pakistan.shtml

Thacker S, Music S, Pollard R, Berggren G, Boulos C, Nagy T, Joseph V (1980) Acute water shortage and health problems in Haiti. The Lancet 315(8166):471–473

The Express Tribune (2018) Pakistan ranks third among countries facing water shortage. Retrieved 28 June 2018, from https://tribune.com.pk/story/1667420/1-pakistan-ranks-third-among-countries-facing-water-shortage/

Wantzen KM, Rothhaupt K-O, Mörtl M, Cantonati M, László G, Fischer P (2008) Ecological effects of water-level fluctuations in lakes: an urgent issue. Hydrobiologia 613(1):1–4

Chapter 10
Socio-economic Impacts of Water Logging in the South-West Coast of Bangladesh

Abdullah Abusayed Khan, Md. Saidur Rashid Sumon and Taufiq-E-Ahmed Shovo

1 Introduction

Over the last decades, the issues related to children and women rights are becoming more prominent, particularly in developing countries such as Bangladesh. Developing countries are faced with the task of protecting their citizenry against a number of calamities in a context of weak governance and unappropriated disaster management systems, inadequate education systems, and underdeveloped coping capacities. In these countries, climate-sensitive health factors such as malnutrition, diarrhoea, and malaria are also high (Haines et al. 2006). South-western Bangladesh, in the vicinity of the Bay of Bengal and Sundarban, has historically been subjected to a plethora of hydro-geomorphological hazards including cyclonic storm surges, floods, soil salinity, and even droughts. Water logging is more recurrent and prolonged here than in other regions of the country. Vulnerability of children in waterlogged areas is thus considered a serious problem in a region where challenges relating to inherent hydrological instability and climate variability can quickly gain extremely grave dimensions. These disasters may have severe consequences for children's physical health and intellectual well-being in both the short and long term (Peek 2008).

Recent and unprecedented water logging calamities have caused extensive damage to roads, bridges, culverts, educational institutions, and dwelling houses, especially thatched and earth-made. Children are particularly affected. Victims of water logging face serious challenges in access to food, shelter, water supply, sanitation, and

A. A. Khan (✉) · T.-E.-A. Shovo
Department of Sociology, Khulna University, Khulna, Bangladesh
e-mail: khanbdnks@yahoo.com

T.-E.-A. Shovo
e-mail: taufiq@soc.ku.ac.bd

Md. S. R. Sumon
Department of Sociology, Jagannath University, Dhaka, Bangladesh
e-mail: saidursumon_du@yahoo.com

© Springer Nature Switzerland AG 2020
S. Bandyopadhyay et al. (eds.), *Water Management in South Asia*,
Contemporary South Asian Studies, https://doi.org/10.1007/978-3-030-35237-0_10

medical facilities. Sources of livelihood are threatened, and the humanitarian crisis persists long after the initial catastrophe. Since the 1980s, the region has been plagued by water logging situations; the recent cyclone *Aila* (devastating storm) made the scenario even worse. The local administration reported that almost a million people were affected in the peak flood time in early August 2011. This particular disaster has unfolded for almost a decade and consists now of a 'persistent disaster'. Increased frequency and unpredictability of these events are, year by year, making communities more vulnerable, particularly women and children. It is also evidenced that some districts in the southern and south-western coastal regions of Bangladesh emerged as new ecologically vulnerable poverty pockets.

Due to perpetual siltation in the rivers and unplanned development interventions on the river system, long-lasting water logging affecting human settlements is taking place in Tala (Satkhira district) resulting in considerable loss and damage to dwelling houses, standing crops, shrimp farms, roads, and educational institutions. Children appear ever more vulnerable due to lack of food, shelter, safe water, sanitation, and medical facilities. People's sources of livelihood became uncertain. Humanitarian situations for affected populations ultimately arise and persist.

2 Purpose of the Study

Issues relating to women and children have been a long-standing phenomenon in south-western Bangladesh; thus far no comprehensive initiatives relating to the water logging situation have been taken. Major objectives in this study include as follows:

- To assess the nature and extent of children and women vulnerability in south-western Bangladesh and proceed with a systematic analysis of the factors involved in this complex phenomenon;
- To identify the socio-structural components and economic factors determining the status of children and women in waterlogged areas in the south-western zone;
- To review and assess the nature and role of both government and NGO provisions regarding child development as well as the degree of implementations in the field.

3 Methodology

For the purpose of the study, a survey method was used which included field observations, documentary/content analysis, focus group discussions, and case studies. This study also uses primary data to address the objectives; meetings and stakeholder consultations selected purposively and proportionately are the main sources for collecting primary data. Data consisting of content analysis were also incorporated: it includes journal, magazine, and newspaper articles and editorials as well as periodic and annual reports regarding the situation of women and children in

Bangladesh, NGOs, and several development partners. Thus, both qualitative and quantitative data have been analysed and interpreted. The unit of analysis consists of children under 18 years old and women aged between 20 and 50+ years old. Jalalpur and the Keshra Union of Tala Upazila in the Satkhira district of Bangladesh were purposively selected for the study. Different vulnerability aspects have been surveyed during focus group discussions; these were then analysed along different dimensions such as food security, health status, housing conditions, educational facilities, social interactions, displacements, agricultural production, drinking water, sanitation, health, and individual life aspirations.

4   Review of Literature

Nearly 40% of Bangladesh's population consists of children; half of these live below the international poverty line. Children and women are facing challenges in areas such as safeguard protection, health, education, nutrition, and safe water and hygiene, all of which are considered the basic rights for all children (UNICEF 2012). The coastal areas of Bangladesh comprise a total area of 42,500 km^2 (32% of the country's total area), with a population of 35 million (around 25% of the total population) distributed along 19 districts (CPD 2000). In 2009, the cyclone *Aila* caused 325 deaths (Jahan 2012), and 103 injuries were reported; over 3.9 million people were affected. Khulna and Shatkhira were the most affected districts with an estimated US$ 270 million worth of asset damages (Roy et al. 2009). Disasters can affect children's personal growth and development (Fuente and Fuentes 2007). Bangladesh is a country with great geographical vulnerability, with no less than 70% of the population living in risk regions. Women, elderly people, and children are most likely to die in natural disasters—whereas, 33% of people sought shelter before disasters, 40% of children did not reach safe shelters and suffered both short-term and long-term consequences including threats to the provision of food, water, health care, education, and communication infrastructure. Women and children experienced most inhuman situations of need and vulnerability as the male earning member of the family either died or migrated so as to make a living CCC (2009). Children and women are not a homogeneous group; different levels and types of vulnerability depend on factors such as age, sex, living standard, ethnicity or tribe, earning or non-earning capacity. The traditional role of children in society is also important; children's positions tend to worsen in disaster situations (Plan International 2005) (Table 1).

4.1   Detailed Discussion of the Situation

Detailed discussion of evidence showed that the people of Jalalpur and Kheshra, particularly women and children, have faced disasters since the very settlement of

Table 1 Summary of focus group discussion

Sl. No.	Date	Place	Participants							Opinion expressed
			Children	Service holder	Business	Farmer	Parents	Teacher	Total	
1	09/08/2015	Jalalpur	1	2	1	2		1	7	*Comments*: **Access to Decision Making:** Children should be considered as an actor to describe the problems
2	09/08/2015	Jalalpur		3	2		2		7	*Comments*: **Compensation**: Local authority should sanction logistics to survival other than needs to compensate them for suffering due to inadequate policy and lack of implementations of projects

(continued)

Table 1 (continued)

Sl. No.	Date	Place	Participants						Total	Opinion expressed
			Children	Service holder	Business	Farmer	Parents	Teacher		
3	11/08/2015	Kheshra				2	3	1	6	*Comments*: In regard to answering the question that who are poor, the FGD participants made for waterlogged vulnerability: Excessive use of land Family belongs to more unemployed members Not able to eat nutritious foods for children Always stay with mental anxiety for child education

(continued)

Table 1 (continued)

| Sl. No. | Date | Place | Participants | | | | | | | Opinion expressed |
			Children	Service holder	Business	Farmer	Parents	Teacher	Total	
4	12/08/2015	Kheshra		1	3		1	3	8	*Comments*: **Socio-economic class and poverty level:** The people who are living on hand to mouth have no specific jobs, worked as servant at others home and landless are considered as poor class. The people who have own cultivable lands above 20 *bighas*, owners of *ghers*, having relatives living in foreign countries considered as rich. All this push them child vulnerabilities

(continued)

Table 1 (continued)

| Sl. No. | Date | Place | Participants | | | | | | | Total | Opinion expressed |
			Children	Service holder	Business	Farmer	Parents	Teacher		
5	12/08/2015	Kheshra			2	1		4	7	***Comments:*** **State of Social Security:** Can move freely without fear during night. Less thief and robbers. Boys and girls cannot move freely ***Suggested:*** Proper compensation has to be given. Opportunity should be given to victims. Some of them wanted rehabilitation

(continued)

Table 1 (continued)

| Sl. No. | Date | Place | Participants | | | | | | | Opinion expressed |
			Children	Service holder	Business	Farmer	Parents	Teacher	Total	
6	13/08/2015	Uttar Shahajadpur, Kheshra		1	3	1		1	6	*Comments*: **Political Stability**: No political commitment yet to be established **Quality of public services**: No good roads (only earthen roads), electricity, drainage system, scarcity of safe drinking water, no waste disposal system, and hygienic toilet for children convenience
8	13/11/2015	Uttar Shahajadpur, Kheshra		3	4	1			8	*Comments*: **Access to information**: Partial access and some cases no scientific communication specifically for children due to lack of electricity facilities
	Total		1	10	15	7	5	8	49	

the village in these locations. Among common disasters, we find floods, cyclones, hailstorms, river erosion, embankment erosion, water logging, and tidal surges. The floods of 1988 and 2009, in particular, hit most areas and caused remarkable damages to the households as well as to cattle and agricultural assets. Cyclones created landfalls and destroyed all houses except concrete buildings. All citizens therefore took refuge in these buildings. For about 15 days, the whole village was under water. Trees and other vegetation were destroyed. The cattle died. All the furniture washed away from the houses. The impact of the damage was essentially the same throughout the area; the succeeding lingering of stagnant water varied from *para* to *para* (one locality to another) (Table 2).

Taking the above data into consideration, the participants of the focus group discussion (FGD) concluded that the Jalalpur village is an area highly disaster-prone area as cyclones, floods, and water logging are all very commonly observed. Most participants presented in the FGD were landless and marginal. Daily income by day labourers was not more than 30 tk. Sometimes, even 'van pullers' fail to find ways to drive as major roads become inundated. In such a situation, people migrate to other areas to earn their livelihoods. Migration is a new technique of survival in both Khesra and Jalalpur as the land-based production system is almost destroyed. Among the 18,730 tube wells in Khesra, 800 were completely inundated during the 2006 water logging disaster. Rainwater is now harvested as drinking water in some areas of Khesra. In these areas, no effective interventions from either the government or NGOs to assist vulnerable populations were found. The water purification system in the area is very poor. Most citizens affected by waterlogging usually drink water directly from ponds, rivers, or other sources carrying strong health risks.

Table 2 Waterlogged situation due to hazard and disasters in case area of Tala Upazila

Year	Hazards	Waterlogged duration (days)	Damages
1969	Cyclone and floods	7–8 days	Mass casualties Cattle perished Damaged houses Damaged trees
1988	Cyclone and floods	7–8 days due to heavy rainfall	Cattle perished Damaged houses Damaged trees
2007	Cyclone and floods	4–6 months	Damaged houses Damaged trees
2009	Cyclones, floods, river erosion, and embankment erosion	2–2.5 months	Cattle perished Damaged houses Damaged trees Saline water invasion
2013	Floods	3–4 days	Damage to crops

4.2   Background Information About the Respondents

The socio-economic and demographic variables of the respondents were meant to consider children vulnerability. These variables are age, gender, religion, family type, family size, employment and education, income, and housing type. Findings of the study reveal that the mean age of the respondents is about 10 years and the highest (48.9%) belong to an age group between 13 and 17 years old. The highest representation (55.1%) is male; 44.9% representation is female. 54.5% of respondents are Muslim and 45.5% Hindu. Most respondents (66.3%) are illiterate; 15.3% have primary education, and 18.4% have secondary or higher education levels. The mean family size is six members. Around 44.9% of respondents live in houses made of thatched, mud, and teen shed; over half respondents (55.5%) has a monthly household income less than BDT. 5000—with an average of BDT 5798. Almost 17.4% of the respondents are day labourers; 12.9% are agricultural workers. 17.4% of respondents are faced with wage discrimination in the workplace, and 5.1% are faced with physical torture, mental stress, and delayed wages.

4.3   Water and Sanitation

Water, sanitation, and hygiene are fundamental human needs, prerequisites to basic human health. About 1.8 million deaths occur every year in coastal areas because of unsafe water and inadequate sanitation and hygiene. Findings reveal that before water logging, 53.9% of respondents collected their drinking water from individual tube wells; during water logging, 74.7% of respondents collected their drinking water from deep community tube wells and 11.2% used rainwater. About 53.7% of respondents have access to sanitation facilities during water logging. Findings also show 36.5% of respondents to have knowledge regarding sanitary latrines; 26.4% have heard about it through word of mouth—9% of respondents have no knowledge about the use of sanitary latrines. Almost 24.2% of respondents do not have access to sanitation facilities due to lack of money to build new latrines and 16.3% due to lack of land space to build latrines. Findings also show that over 59.0% of respondents lack access to water as well as sanitation and hygienic facilities; as a result, 43.8% suffer from diarrhoea, and 10.1% suffer from scabies.

4.4   Diseases and Medical Facilities

The waterlogged areas of Bangladesh are highly vulnerable to outbreaks of infectious, waterborne, and other types of diseases. A number of waterborne diseases, including diarrhoea, dysentery, cholera, and skin diseases, etc., are very common in Bangladesh. The lack of safe drinking water may be the most important cause for the spread of waterborne diseases during and after conditions of water logging. 75.3% of respondents were affected by diarrhoea and 59.6% by scabies. The most widespread disease is diarrhoea (25.3% of respondents); scabies follows with 24.2% of respondents. The majority of respondents (46.1%) attends professional *allopathic* doctors for treatment; 16.9% resort to self-care. About 29.8% of respondents visited a public hospital, 24.2% visited local doctors, 18.5% visited the community clinic, and 12.9% visited the local pharmacy.

4.5 *Various Dimensions of Vulnerability*

Findings reveal that 56.2% of respondents received information about water logging disaster from word of mouth. 23.0% received information through the TV and 12.9% through the local mosque, a cell phone, from a disaster control centre or through radio transmission. During the water logging itself, many respondents (39.3%) took refuge in nearby buildings, strong houses, school buildings, and mosques; 25.3% went to a nearby safe zone. In this period, 87.1% of respondents faced food shortages, and 29.8% had access to only irregular meals. 87.1% of respondents faced economic problems, and 66.3% faced problems of food scarcity. During the economic crisis, 65.2% of respondents received monetary support from local NGOs' and institutions. 42.1% of respondents resorted to moneylenders (Fig. 1).

5 Discussion of the Results

About 34% of respondents changed their sources of livelihood due to water logging conditions. Most mud-built houses were destroyed, and almost 50% of FGD participants reported to be living in refugee centres with their children. Respondents also complained that inadequate transportation systems are a major cause of high school dropout rates. Preservation of educational materials is a difficult task under waterlogged conditions. Thirty-nine per cent of respondents received medical treatment from a public hospital. Sixty-three per cent are affected by diarrhoea and cold, and 72% lack access to purified water. FGD with women and children revealed that, in many cases, people are not keen to establish marital relationships with women from areas affected by water logging as these women generally suffer from skin diseases. Sanitation status is a major indicator for assessing women vulnerability. Almost 36% of women go secretly to dry lands, jungle areas, or open fields for defecation. It was found that 30% use *kutcha* (which is made by bamboo or wood and consists of a feeble structure) *latrines,* and 46% defecate in the pond/river. Only 2% use *pucca* (comparatively strong structure, usually made by concrete) latrines; others (9%) defecate directly at home so that nobody can see them during daylight. At night, they throw the litters into nearby ponds or jungle areas. In these case areas, no effective interventions exist to assist vulnerable people from either government or NGOs.

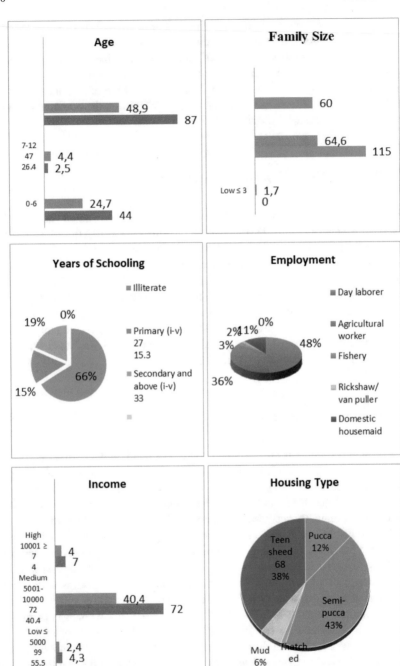

Fig. 1 Characteristics of the surveyed persons

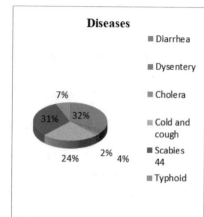

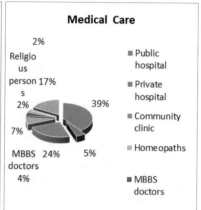

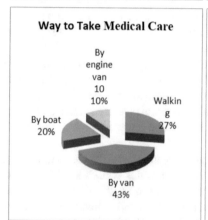

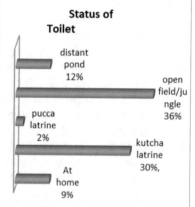

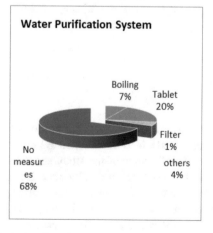

Fig. 1 (continued)

5.1 Policy Recommendations

On the basis of the above findings and analysis, the following recommendations can be put forward for further intervention:

(a) Initiatives may be taken to develop regional funds for addressing issues regarding vulnerabilities affecting women and children. These initiatives will help affected areas overcome immediate risks so that children can resume their normal lives as soon as possible.

(b) Developing regional and international cooperation in the form of meetings, workshops, or study tours aiming to articulate a common understanding of problems relating to women and children. We also need to address issues such as child labour and vulnerabilities in waterlogged areas as well as their ramifications.

(c) The first initiatives should be taken by organizations such as the government and NGOs so as to create a safe, healthy, and respectful environment and give workers the opportunities to ask questions and express concerns about the prevention of child abuse, particularly girl abuse.

(d) A comprehensive package of compensations should be offered to both landowners and land users in connection with children's vulnerability so as to improve children and women livelihoods and reduce suffering in these waterlogged areas.

(e) Social awareness of the needs of local communities is needed when developing tidal river management. Intensive consultation and negotiation with affected communities are crucial.

(f) We need strong advocacy by civil rights organizations, NGOs, and local government authorities so as to discuss relevant issues at both local and national levels. Improving livelihoods and providing both short- and long-term health safety and life insurance are key concerns. Legislative interventions are also needed that consider conditions in the field. Effective interventions demand comprehensive cooperation across hierarchies, between agencies and across geographically based bureaucratic structures.

6 Conclusion

Due to time constraints, the present study does not cover a large proportion of the waterlogged study areas (in particular Tala Upzila in south-western Bangladesh) where women and children are vulnerable. The findings of the study showed that children and women in these areas are severely exploited and vulnerable in social, economic, and cultural spheres. Poor infrastructure and traditional tools and techniques affect people negatively, particularly children and women. Tidal approaches such as the tidal river management (TRM) in the broader spectrum seem to be an effective and sustainable way forward. Evidently, the field study portrayed certain low-laying areas such as the Jalalpur and Kheshra unions as moderately high-tide-prone areas, thus naturally helping tidal channels deepening. From the local point of view, most people showed a keen interest in the implementation of tidal management and more scientific knowledge for disaster management. In this connection, tidal river management with the help of technology, based sandy dumping on the river or water basin development, seems to be the most feasible and effective options for disaster management. More research is needed, particularly as regards women and children's livelihoods through lingering waterlogged situations in affected areas.

References

CCC (2009) Climate change and health impacts in Bangladesh. Climate Change Cell, DoE, MoEF; Component 4b, CDMP, MoFDM, Dhaka

CPD (2000) Female headed households in rural Bangladesh: strategies for wellbeing and survival, Centre for Policy Dialogue (CPD), World Bank

Fuente AD, Fuentes R (2007) The impact of natural disasters on children morbidity in rural Mexico. HDR Occasional paper 2007/4, New York

Haines A, Kovats RS, Campbell-Lendrum D, Corvalan C (2006) Climate change and human health: impacts, vulnerability and public health. Public Health 120:585–596

Jahan I (2012) Cyclone Aila and the Southwestern Coastal Zone of Bangladesh: in the context of vulnerability. Master thesis, Lund University

Peek L (2008) Children and disasters: understanding vulnerability, developing capacities, and promoting resilience—an introduction. Child Youth Environ 18(1):1–29

Plan International (2005) After the cameras have gone: children in disasters. Plan International London

Roy K, Kumar U, Mehedi H, Sultana T, Ershad DM (2009) Cyclone Aila 25 May 2009. Unnayan Onneshan-The Innovators, Dhaka

UNICEF (2012) UNICEF and child-centred disaster risk reduction. United Nations Children's Fund

Chapter 11
India–Pakistan River Water Sharing: Prospects for Environmental Peacebuilding

Dhanasree Jayaram

1 Introduction

Bitter political rivalry between India and Pakistan has led the two countries to fight three full-fledged and one limited war (under the nuclear shadow) since Independence from the British Empire in 1947. As geopolitical and territorial tensions over the Kashmir dispute and cross-border terrorism continue to simmer, relations between the two countries in other sectors such as trade, the environment, culture, education and so on have much suffered. Such realities have also complicated India–Pakistan water-sharing relations to such an extent that the two countries, despite signing the Indus Waters Treaty (IWT) in 1960, have not been able to translate it into any form of "effective" cooperation capable of addressing increasing water insecurity issues within the Indus Basin. In fact, the treaty has become a mere tool for division of waters and dispute resolution. The most recent incident in this area was the 2016 Uri attack on the Indian Army brigade, a cross-border attack traced to a Pakistan-based terrorist outfit—and after which the Indian Prime Minister stated that blood and water could not flow together.

From the time, water was apportioned by the Inter-Dominion Accord in 1948 up to the signing of the IWT in 1960, India and Pakistan engaged in negotiations for a long-lasting treaty. However, the treaty is at best a sub-optimal agreement that ensuring that the rivers were divided between the two countries—without much scope for true, functional cooperation over water management. The IWT was brokered by the World Bank and signed by Prime Minister of India, Jawaharlal Nehru and President of Pakistan, Mohammad Ayub Khan. There are several aspects of this treaty worthy of analysis—including its constitutionality, since the treaty was not signed by the Indian President and Head of State but only by the Prime Minister as mere Head of Government. That issue was settled after India's Supreme Court rejected a Public

D. Jayaram (✉)
Department of Geopolitics and International Relations, Centre for Climate Studies, Manipal Academy of Higher Education (MAHE), Manipal, Karnataka, India
e-mail: dhanasree.j@manipal.edu

© Springer Nature Switzerland AG 2020
S. Bandyopadhyay et al. (eds.), *Water Management in South Asia*,
Contemporary South Asian Studies, https://doi.org/10.1007/978-3-030-35237-0_11

Interest Litigation (PIL) that sought to declare the treaty as unconstitutional (PTI 2017). Similarly, the role of the World Bank, which brokered the treaty, has come under scrutiny. Furthermore, for different reasons, neither India nor Pakistan have acceded to the UN Convention on the Law of Non-Navigational Uses of International Watercourses.

There have been many disputes relating to shared waters, especially over the construction of dams on the Indian side of the border (such as the Baglihar Dam on the River Chenab and the Kishanganga project on the River Jhelum) and the demand by India for a greater share of available water (it is less than 20% under the IWT). Climate change is also exacerbating the situation, which can only worsen in the future as the rivers of the Indus Basin are dependent on glacial meltwater and monsoons, and two factors heavily influenced by climatic changes. Furthermore, politico-security issues such as the role of the distribution of the Indus Basin's waters in the Kashmir dispute and the threat by Pakistan to use nuclear weapons against India if the latter chokes water supplies have escalated a resource-related conflict to the level of a potential water war. However, one must remember that despite political and military antagonism, the principles of the IWT have been adhered to by both countries over the decades—and even in wartime. The Indus (and its tributaries) constitute the lifeline of Pakistan. Of the population that is supported by the Indus basin, whether for agricultural purposes or drinking water, 61% live in Pakistan and 35% in India–Pakistan being thus much more dependent and vulnerable than India (Roic et al. 2017: 59). However, on the Indian side as well, with decreasing availability of water per capita, there is a call for boosting the use of the rivers on which, although not allocated to India, it is allowed to build infrastructure for non-consumptive purposes.

Nevertheless, the challenges facing both India and Pakistan are immense, especially as the effects of environmental change are felt more and more severely on both sides of the border. Instead of revising the treaty and wrestling over riparian politics, India and Pakistan could work towards integrated basin management as water security is a critical issue in both countries. Groundwater management in border regions was left out of the negotiations even though the aquifers are shared and are fast depleting, especially in the Punjab provinces on both sides of the border. Water management negotiations are the biggest confidence-building and peacebuilding measure, which could potentially transform relations between the two countries in other sectors too. In this context, the paper traces the history of river water-sharing relations between India and Pakistan amidst geopolitical and military hostilities. It throws light on conflict and cooperation between the two nations over the distribution and management of the waters of the Indus Basin by using the theoretical debate between "environmental conflict" and "environmental cooperation"—primarily, the conceptual framework of "environmental peacebuilding" in the context of water relations. It problematises the concept of cooperation not by juxtaposing it with conflict but by delineating different gradations of cooperation and highlighting "effective" cooperation as a model for environmental peacebuilding. Conclusions are arrived at by citing different documents, including the United Nations' toolkit for environmental peacebuilding. The paper concludes by exploring possible starting points for water cooperation between

India and Pakistan (such as integrated basin management) which could pave the way for environmental peacebuilding between the two countries in the long term.

2   Environmental Cooperation in the Environmental Security Discourse

Within the environmental security discourse, theses connecting environmental scarcity to conflict have had an upper hand in comparison with those focusing on environmental cooperation. One of the rationales for the need to develop thesis emphasising environment-conflict situations has been to raise the urgency of environmental concerns that contribute to civil wars, conflicts and instability and to reiterate the relevance of integrating environmental variables in the international and/or regional security-related architectures as well as national security strategies. Calls by various scholars such as Ullman (1983), Romm (1993), Myers (1996), Matthews (1989) and Kaplan (1994) for a widening and deepening of the "security" realm by incorporating issues such as environmental change led to a movement towards redefining security—however, as already stated, with a great emphasis on policy alone. From an "interdisciplinary debate" involving members of the academic and political communities—regarding whether or not environmental issues should be integrated into security concerns, and if so, how should they be incorporated—the discourse moved towards "process tracing" and empirical research in the 1990s. In the attempt to build the empirical foundations for environmental security research, scholars such as Homer-Dixon (1999) and Baechler (1998) investigated the links between resource scarcities, particularly renewable ones such as water, and violent conflict. Rather than dealing with a large gamut of environmental challenges, these studies dealt with a few localised issues such as cropland, fish stocks, forest and water; also, instead of going too much deep into security at the conceptual level, they focused upon acute conflict—at a national or international level. It must also be pointed out that resource abundance, whether it may be that of agricultural land, food, forests, fisheries or water, has also been a large concern in the environmental security discourse. Transboundary conflicts, specifically wherein natural resources are involved, is one among several types of conflicts characterised as environmental conflicts. Yet another factor in the environment-conflict thesis is the connection between population growth and violent conflict, based on increased consumption and degradation of resources which leads to the lack either of physical availability of resources or lack of access (in short, scarcity). The majority of environmental security scholars have concluded that, whereas environmental change is not an immediate cause for conflict, it can certainly act as a threat multiplier—as an exacerbating factor. As environmentally induced international wars between two countries are highly unlikely, most studies concentrate mainly on sub-national conflicts.

Although the environmental security discourse has mostly focuses on environment-conflict thesis, there is an increasing momentum towards focusing on transboundary environmental management and cooperation. Critics are of the former thesis are emerging who question the logic of countries waging wars over resources that are existential in nature and very hard to monopolise due to their transboundary characteristics—which by their own nature dictate a collaborative approach. The critique against conflicts over environmental and resource scarcities is premised upon the rationale that by waging a war against your adversary, there is no guarantee that you would be in a position to ensure long-term access to the resource in contention; in fact, the probability of diminution of resource access is greater due to various intervening factors such as degradation and pollution. According to Conca and Dabelko (2002), Barnett (2000) and Deudney (1990), countries are more inclined (indeed obliged) to cooperate as regards common environmental challenges. However, this does not imply that the mere existence of common environmental challenges always leads to environmental cooperation. According to the literature on environmental cooperation, for countries to reach a consensus on signing an agreement or creating a cooperative regime related to transboundary environmental cooperation, besides common environmental challenges there need to be further common interests as well. If the concerned nation states have competing interests in relation to the environmental challenge, cooperation would be difficult to materialise. At the same time, scholars such as Dabelko and Conca emphasise the importance of shifting attention from environmental causes of conflict (as is mostly the case with environment-conflict thesis) to "cooperative triggers of peace that shared environmental problems might make available" (Conca and Dabelko 2002: 5). Similarly, there is a section of the literature concerning environmental cooperation that describes it as a confidence-building tool as well as a means for building awareness in terms of finding common or shared solutions to common problems (Maas et al. 2013).

The environmental cooperation discourse is founded upon various propositions, the primary among which is that environmental security can best be ensured through cooperation rather than through war and associated destruction [of resources]. When seen through the prism of neorealism, in an anarchical international system, environment could be considered a "soft" issue regarding which countries could easily cooperate. This system is influenced by factors such as globalisation, interdependence and institutionalisation, which then prompt nation states to cooperate and resolve problems jointly. This can be seen in the context of ideas such as "collective security" which is premised upon state security achieved through international, collective action (Claude 1956); or even in the context of international regimes that are usually associated with neoliberalism. However, as argued by Axelrod and Keohane (1986), cooperation is possible only when there is scope for "absolute gain" or "mutual gain", as opposed to a "zero-sum game" or "relative gain". From a realist perspective, especially defensive structural realism, cooperation is hindered by the "security dilemma" as states strive to increase their own security in the context of international anarchy, leading inadvertently to decreased security for others and provoking these to heighten their own capabilities. However, defensive realists point towards ways in which the "security dilemma" can be reduced by increasing the gains of cooperation as well as the costs of non-cooperation (Jervis 1978). Moreover, in the context of environmental cooperation, it can be argued that since the threats and/or risks posed by environmental change and/or disruption are most often "common"—or "shared"—cooperation is a more viable, (and at times simply necessary) option, despite the fact that the capacity to adapt may not be the same among all involved. What may also influence decision-making on environmental cooperation is the fact that the scientific, technical and logistical information regarding such changes and disruptions is relatively common—and, more importantly, unclassified. Hence, it could be contended that parties—including nation states—even when politically and diplomatically hostile to each other, are likelier to cooperate as regards environmental challenges. The most oft-quoted examples are that of river water sharing in the Middle East or West and South Asia.

At the same time, even while making the case for environmental cooperation, scholars have reflected upon the barriers involved, some of which have already been mentioned. When it comes to interstate environmental challenges, nation states are largely the only legitimate actors in a position of choosing to cooperate or not. In such a scenario, ideas of sovereignty and national interests still factor in a big way into the cooperative arrangements undertaken and the negotiations that lead to such arrangements. For instance, a given state's perception of the costs involved in cooperating will guide its approach to negotiations. This is why many negotiations fail at the outset; sometimes initiatives that are targeted at environmental gains do not translate into a formal codified agreement; even when agreements are reached, they are often either not legally binding or constitute a sub-optimal agreement (Conca and Dabelko 2002: 9). Similarly, although neorealism acknowledges environmental cooperation, it adopts a worldview based on factors such as environmental vulnerability (which varies from nation state to nation state), cost–benefit analysis carried out by states, capacity by nation states (which again is variable in an "anarchical international system") and willingness among nation states (Henriette 2012: 59).

It is therefore not enough for states to share common problems for them to cooperate bilaterally and/or multilaterally—convergence of interests and views among nation states do not necessarily translate into effective cooperation. The literature concerning environmental cooperation lists out many obstacles either preventing a cooperative arrangement or impeding its progress/implementation after institutionalisation. First and foremost, environmental challenges are highly complex and most often interconnected with political, socio-economic and security-related issues. Therefore, in many cases, trade-offs are an essential part of negotiations and decision-making. Very often, security and political drivers leaning towards non-cooperation trump environmental factors, resulting in non-cooperation. Second, even though environmental agreements are reached between nation states (at the interstate level), the stakeholders involved are many and the agreement has an impact on the larger civil society, non-governmental organisations and corporations. Third, environmental problems are usually associated with uncertainty and unpredictability, which again tends to delay any form of environmental cooperation. As argued by scholars such as Sjöstedt (2009), most environmental problems are categorised as risks and environmental cooperation is known to materialise mainly only when a crisis emanates.

Fourth, what is equally important to emphasise is that the mere signing of an agreement may not qualify as cooperation; it is rather the actual content of cooperation that defines environmental cooperation (Conca and Dabelko 2002). This is particularly true in many instances of river water-sharing agreements, including the Indus Waters Treaty, which are based on division of waters rather than integrated water resource management or similar joint initiatives. Also, cooperation is dictated by existing levels of knowledge among concerned stakeholders; when information is insufficient, agents fail to design or implement an action plan in the most effective manner. In addition, domestic conditions also have a bearing on the outcome of such cooperative arrangements, as some nation states are more willing and have more resources to initiate and implement action plans than others. Fifth, even while converging on the interests of cooperation, concerned stakeholders could espouse completely or partially diverging positions in the solution-finding exercise, especially when the costs and benefits of a solution are distributed asymmetrically among stakeholders. Sixth, solutions to environmental challenges normally follow a very wide timeframe due to their complexity—and therefore the benefits of cooperation are realised mostly in the long term while the costs involved in adopting the solution are borne in the short term. When short-term costs are high, states may choose not to cooperate, or at best delay cooperation. It is, however, not necessary to view these drivers as mere obstacles; they can be endure some unexpected twists as well. For example, scholars have argued that the greater the level of uncertainty, the greater the chances of nation states or other stakeholders choosing to cooperate, as they would be wary of the potential costs of non-cooperation (Henriette 2012).

3 Environmental Cooperation for Peacebuilding

The United Nations Environment Programme (UNEP) founded a programme on "Environmental Cooperation for Peacebuilding" (ECP) in 2008, with the overall aim of "strengthen[ing] the capacity of countries, regional organisations, UN entities and civil society to understand and respond to the conflict risks and peacebuilding opportunities presented by natural resources and environment" (United Nations Environment Programme 2015: 8). Similarly, the Environmental Peacebuilding Association (EnPAx) defines "environmental peacebuilding" as an activity that "integrates natural resource management in conflict prevention, mitigation, resolution and recovery to build resilience in communities affected by conflict". Although the focus of the UN and forums such as the EnPAx has generally been on post-conflict scenarios, environmental peacebuilding is increasingly being used in pre-conflict scenarios and intractable conflicts—especially in the context of sharing of natural resources between two or more countries which are hostile towards each other politically, diplomatically or simply hostile regarding the sharing of a natural resource. At the interstate level, environmental peacebuilding can largely be considered a subset of "environmental cooperation", wherein the focus is on cooperation on environmental issues—whether or not environmental factors caused, triggered or exacerbated the conflict itself—and wherein environmental cooperation is used as a tool for peacebuilding. As opposed to the environment-conflict thesis, environmental peace-building does not build its premise on "environmental causality in political conflicts, such as civil war" in order to "accentuate the importance of conservation". Instead, it deems environmental issues as pertinent entry points for engagement, and even cooperation, in other realms, including the political realm. Environmental concerns can also emerge as a "valuable exit strategy from intractable deadlocks" between political and diplomatic adversaries. Technical cooperation—which cannot neces-sarily be delinked from other factors—has the potential to develop a certain level of trust among nation states through knowledge-sharing and other mechanisms, which in turn could engender diplomatic initiatives, leading to "peace dividends". Envi-ronmental peacebuilding does not entirely depend on common interests, as often common interests can even result in competition and adversarial behaviour. Instead, it draws principally on the assumption that common environmental threats induce cooperation and peacebuilding among nation states (Ali and Vladich 2016: 610–11).

At the same time, environmental peacebuilding has been criticised on many fronts. While some argue that it can never be a part of high politics or translate into coop-eration in high politics, others express reservations about its impact owing to the timescale of the processes involved. Usually, the results of environmental peace-building fructify over a long period of time as they involve a series of steps, including the creation of a knowledge database, recognition of the need to act upon a common environmental threat through cooperative mechanisms, trust-building through envi-ronmental cooperation, continued communication owing to the necessity or existen-tial nature of the problem, gradual attenuation of misunderstandings and other bones of contention, conflict de-escalation and ultimately, peacebuilding (Ali and Vladich 2016: 611–12). Peace ecologists have also critiqued environmental peacebuilding on the basis that the latter depends on environmental problem-solving, referring to categorisation of environmental peacebuilding initiatives into "efforts to prevent con-flicts related directly to the environment"; "attempts to initiate and maintain dialogue between parties in conflict"; and "initiatives to create a sustainable basis for peace" (Conca et al. 2005: 150). According to this view, environmental peace-making or peacebuilding should be defined as "the identification and utilisation of opportunities from the natural and human environment for building bridges of communication and collaboration among parties in conflict". This definition goes beyond the functional aspects associated with specific projects undertaken for environmental peacebuilding to integrate different values such as interconnectedness and interdependence between biodiversity and cultural diversity, "bioregionalism" and a much broader definition of environment that encompasses "sense of place, ecological values and cultural–ecological heritage" among others. It does not depend on an environmental problem but takes into account the "values, interests, worldviews, ideologies and theologies relating to the environment" (Kyrou 2007).

Many scholars have also investigated the role of environmental peacebuilding in the water sector, whether transboundary or sub-national. Lowi (1993), for instance, analyses prospects for cooperation between disputing nations in transboundary river basins by employing a framework that takes into account the following variables:

"resource need/dependence", "relative power resources", "character of riparian relations or the impact of the larger political conflict" and "efforts at conflict resolution a third-party involvement". Through these variables, Lowi adopts a combination of realist and constructivist approaches so as to study concerns that have influenced interstate behaviour in a river basin, albeit her analysis is predominantly "realist". In this framework, both capabilities or resources (economic, military, military, etc.) and core values and concerns regarding identity and legitimacy are influential in framing water cooperation dynamics between nation states. Aggestam and Sundell-Eklund (2014) discuss the linkage between hydro-politics and peacebuilding, with the former referred to as "conflicts and negotiation processes between sovereign states on water allocation and distribution, particularly in relation to transboundary rivers or aquifers". In contrast to functionalist theory, which propagates that it is possible to nurture cooperative relations in areas of low politics, that could have spillover effects in areas of high politics, as also envisaged in the environmental peacebuilding discourse, these scholars uphold the criticality of rights-based framing of water negotiations. These functionalist assumptions also form a major part of negotiation theory, based on the "principle of gradualism" (Weiss 2003). Gradualism is typically known for moving from simpler issues to more complex ones. As in the beginning the degree of trust between the conflictual parties is considerably low, it is both practical and more feasible to start with smaller issues and create a sense of ease among the parties. This approach has been borrowed by the environmental peacebuilding discourse as well, particularly with relation to water cooperation.

From a purely functionalist perspective, water diplomacy is one of the prominent frameworks dealing with prevention and resolution of water conflicts and facilitation of water cooperation. Water diplomacy includes "all measures by state and non-state actors that can be undertaken to prevent or peacefully resolve (emerging) conflicts and facilitate cooperation related to water availability, allocation or use between and within states and public and private stakeholders". The water diplomacy framework looks beyond the narrow scope of technological or values-based approaches and follows a "know-why" perspective, rather than a "know-how" one, which influences the choice of tools—with adequate consideration of how values and interests define the water conflict or water problem in question. It "diagnoses water problems, identifies intervention points and proposes sustainable resolutions that incorporate diverse viewpoints and uncertainty as well as changing and competing demands" (Islam and Repella 2015: 1). This approach takes into account social, economic, political and ethical and other such considerations, for sustainability of both solutions and tools. For example, there are various values associated with the Indus that are of consequence to both India and Pakistan—territoriality, human rights, sovereignty, and economic productivity and so on. This framework uses negotiation as the bedrock for cooperation and thus peacebuilding. It highlights the relevance of identifying the "zone of possible effective cooperation" (ZOPEC)—and not just the "zone of possible agreement" (ZOPA), more common in negotiation literature. Effective cooperation is therefore defined as "a collaboration in which two or more parties find a negotiated compromise on maximising mutual gains and achieving joint wins for all parties involved, resulting in the availability of acceptable quantity and quality of water for health, livelihoods, ecosystems and production, coupled with an acceptable level of water-related risks to people, environments and economies" (Huntjens et al. 2016: 50). In the case of the Indus basin, India and Pakistan do not seem to be engaging in "effective cooperation", as both countries cry foul over each other's intentions, attitudes and activities.

Effective cooperation is also at the heart of several analyses concerning international water cooperation, and this can be seen in the case of the Indus basin. Very often it is the absence of violence or peace defined in negative terms—rather than positive peace—that characterise water cooperation; yet it is positive peace which is seen as a requisite for lasting, peaceful interstate relations (environmental peacebuilding) as well as long-term environmental security ensuring the protection of the environment, citizens and states. At the same time, studies have shown that in recent times, there has been an increase in the number of "positive, water-related interactions" between states, especially those "not in acute conflict with each other" (Ide and Detges 2018: 63–84). From a practitioner's point of view, it has been argued that interactions among states cannot be regarded as "cooperation" when they do not produce any measurable benefits. Measurable benefits—either quantifiable (such as shared revenues from hydropower) or unquantifiable (such as lessened tensions between countries)—which address water stress/scarcity and improve water security are vital for effective cooperation. Hence, it is not enough to look at the processes and practices by nation states; one must also consider results so as to measure effective cooperation (Tarlock 2015). Moreover, while a large part of the literature focuses on water scarcity alone when analysing environmental conflict and cooperation, there is a need to delve into wider issues such as historical and socio-economic relations between the concerned nation states, the nature of the decision-making regimes and other political and bureaucratic institutions and the level of economic development. Therefore, Dinar (2011: 172) argues that, in the context of water conflict and cooperation, it is equally important to take into account "geographic characteristics of a river basin and the location of the riparian countries along the river, the relative distribution of power among the basin's riparians, and the domestic political arenas in the respective countries".

The literature concerning water cooperation focuses to a large extent on the benefits of cooperation, which range from "environmental benefits to the river" to "economic benefits from the river", "reduction of costs because of the river" and "benefits beyond the river" (Sadoff and Grey 2002). Although environmental peacebuilding targets all four benefits, what it mainly aims at are the last two. On the one hand, it strives to reduce geopolitical tensions between states; on the other, it aims to catalyse cooperation in other political and socio-economic sectors. The UN's environmental peacebuilding toolkit also lists indicators of cooperation, one of which is termed "benefit achievement" (technical, social, economic and political and can be seen as another indicator of effective cooperation. The intensity of cooperation between conflicting parties can be measured through eight other indicators: "commonality and shared purpose", "institutional change", "multiplexity and scope", "continuity of engagement", "information transparency", "cooperation intensity", "dispute resolution capacity" and "stakeholder involvement". The intensity of cooperation is lowest when only technical benefits are achieved. This state of affairs is followed by a combination of technical and social benefits; further, it is achieved a combination of technical, social and economic benefits; cooperation is highest when benefits include technical, social, economic and political blessings. For instance, India–Pakistan water cooperation is typically characterised by level I in most indicators and by level II and level III in a few areas such as information sharing (which is institutionalised in this case) and resolution process (with the signing of the Indus Waters Treaty and continual of the process in successive disputes). As a juxtaposition to cooperation, the UN document also enumerates indicators of increasing conflict intensity: "conflict analysis and programming", "level of politicisation" and "applied power". In the case of India–Pakistan water relations, the intensity of conflict can be gauged by the two countries' constant "strong verbal threats" or "ultimatums" and/or "use of coercive instruments or display of symbolic acts". According to the scale proposed by the UN's toolkit, the intensity of conflict would be level II. Level IV is representative of parties that indulge in violence, military campaigns or resource capture (United Nations Interagency Framework Team for Preventive Action 2012: 104–105). Further, analysis of India–Pakistan water relations will be carried out in the subsequent sections.

It must also be noted here that although a large section of the literature deals with water cooperation as a normative framework, Zeitoun and Mirumachi (2008) and Zeitoun and Warner (2006) have contended that not all conflicts related to water should be seen through a negative lens. In the same vein, not all cooperative arrangements are inherently positive, as some conflicts can lead to intrinsic disputes being addressed and cooperative agreements may reflect and even fortify "power inequities" and "disparities"—a phenomenon referred to as "hydro-hegemony". Hydro-hegemony is described thus: "the most powerful country in the basin, the hydro-hegemon, can create its preferred mechanisms of transboundary water management due to its relative power within the watershed". Due to its higher structural power, the more powerful country in the basin is capable of enforcing asymmetric and inequitable cooperation dynamics (Petersen-Perlman et al. 2017). Similarly, desecuritisation of water conflicts, coupled with the use of "functional technocracy", has also been seen with scepticism as the common principles involved in a technocracy such as "professionalism, standardisation and rational problem-solving" could cloud "hydro-politics and power dynamics", which in turn would "strengthen the status quo rather than resolving the conflict". It also leads to marginalisation of certain actors as the process is dominated by "water experts and development brokers" who are "assumed to be impartial and unbiased on the basis of their technical and/or scientific knowledge and competence". For instance, the case of Israel–Palestine cooperation reflects to a certain extent the negative consequences of discretisation and overemphasis on technocracy, wherein Israel's framing of water issues has dominated and many disempowered actors within Palestine were side-lined in the decision-making process (Aggestam 2015).

4   The Indus Waters Treaty: "Water Cooperation" Between India and Pakistan?

The conflict between India and Pakistan over the waters of the Indus basin began to brew right after Partition in 1948. Although initially a short-term Inter-Dominion Accord was signed so as to apportion the waters in such a way that India would release sufficient amount of water to Pakistan in return for annual payments, by 1951, the situation worsened: Pakistan demanded that the matter be taken to the International Court of Justice; India refused to do so and opted for a bilateral resolution. It goes without saying that the partition left an eternal scar on India–Pakistan relations and tensions between the two countries had heightened over the Kashmir dispute as well. It was at this juncture that David Lilienthal, former chairman of the Tennessee Valley Authority and the United States Atomic Energy Commission, visited both India and Pakistan. He recommended that the World Bank should act as a mediator, helping the two countries come to an agreement on how to jointly manage the waters of the Indus Basin river system—and utilise, in the best possible manner, the plentiful water resources through dams and irrigation canals so as to boost food production. With the Bank agreeing to act as a third-party arbitrate, India decided to come on board on the condition that the Bank would not act as an adjudicator and would only facilitate the negotiations while letting the differences be resolved bilaterally.

Right from the beginning, it was clear that neither India nor Pakistan would compromise much on their positions—especially since Pakistan hinged its entire negotiating position on its historical right to the waters of the Indus basin, while India did not want historical distribution of waters to dictate future allocation. It must also be noted that immediately after the partition, the governments of both west and east Punjab (in Pakistan and India, respectively) had signed a few temporary agreements with regard to the distribution of water and so as to maintain water supply to Pakistan. However, when these agreements elapsed in March 1948, east Punjab is known to have cut off water supply to west Punjab, triggering a sense of weakness and fear in Pakistan. This realisation that it could face water shortages—and that it would always be at a disadvantage when negotiating with India regarding water due its to lower riparian position—heavily contributed to Pakistan's stance on the subsequent initiatives intended to resolve the water dispute between the two countries (Bhatnagar 2014: 273–274). Nevertheless, the World Bank played a critical role in designing a proposal for negotiations, which was based on three principles: first, the water resources in the Indus basin are adequate to meet both the current and future needs of both India and Pakistan; second, the Indus basin should be treated as a unit and the water resources of the basin should be harnessed in a cooperative manner; and third, the resolution should draw upon functional principles, far removed from political considerations that could potentially stymie cooperation. The Bank's move to treat the matter as an engineering problem that could be fixed through technocratic solutions did not achieve much success as politics superseded the technical aspects of distribution or allocation. Nevertheless, both countries were aware of the fact that a resolution to the Indus dispute was a necessity for avoiding a future conflict over water as well as to proceed with several river development projects. After the 1948 Inter-Dominion Agreement failed to bring any resolution to disputes such as that relating to the construction of irrigation projects (especially by India) on the River Sutlej, the World Bank announced that unless an agreement was reached, the Bank would not be in a position to finance these or other development projects which both India and Pakistan were in need of and had applied for. Both countries thus agreed to the World Bank's mediatory role in 1951 and—after nine long years of negotiations—the IWT was reached in 1960 (Miner et al. 2009: 205–207).

Under the IWT, all the waters of the eastern rivers (Sutlej, Beas and Ravi) have been allocated to India for unrestricted use while Pakistan would be allowed to use these waters only for domestic and non-consumptive uses—with no interference with the flow until it crossed over into its territory. However, during the transition period, lasting from 1 April 1960 to 31 March 1970, India would limit the use of these waters for storage and agricultural use as well as make deliveries to Pakistan. Similarly, the western rivers (Indus, Chenab and Jhelum) have been allocated to Pakistan. The World Bank, which is also a signatory to several provisions of the treaty, is in charge of operating the Indus Basin Development Fund—as well as conflict resolution mechanisms under Articles V and X as well as Annexures F, G and H. However, the Bank effectively withdrew from its initial role in 1970, leaving the operationalisation and implementation of the treaty to the Permanent Indus Commission, consisting of a commissioner—usually a "high-ranking engineer competent in the field of hydrology and water-use" (The Indus Waters Treaty 1960)— from each country so as to maintain a regular channel for communication and address questions related to the treaty's implementation (Table 1).

In this context, while several frameworks and approaches have been used to analyse India–Pakistan water-sharing relations, two stand out—"water wars rationale" and "water rationality". The water wars' rationale has been used extensively by advocates of the environment-conflict thesis to establish the linkages between environmental or resource scarcity and conflict. This rationale predicts that the two countries are likely to engage in conflict over access to limited supplies of water. Gleick (1991) has summarised several ways in which water could play a significant role in conflicts and wars, three of which are especially relevant to the case of India–Pakistan. First, water resources could act as "military and strategic goals", on account of which countries could go to war so as to ensure access to water or seize water resources. This is based on several factors such as the "degree of scarcity", "the extent to which

Table 1 India and Pakistan's shares of the Indus River system waters (mean yearly flows)

Western rivers	Eastern rivers	Total Indus system flows
167.2 BCM	40.4 BCM	206.7 BCM
Pakistan's share of total Indus system rivers: 80.52%	India's share of total Indus system: 19.48%	100%

Source Chellaney (2011)
Notes BCM = billion cubic metres

the water supply is shared by more than one region or state", "the relative power of the basin states" and "the ease of access to alternative freshwater resources". Second, during a military-led campaign, water systems can be used as "both the targets and the tools of war". It has been seen in the past that during wars, water-resource infrastructures such as dams, canals and desalination plants have been attacked. Third, the likelihood of conflicts and disputes due to resource inequities both within and between the concerned countries is significant as differences over irrigation, hydro-electric, flood control and other such projects emerge or widen. Similarly, many empirical studies have also highlighted the interconnection between water scarcity and armed conflict or militarised disputes (Gleditsch 1998) (Hensel et al. 2006). Some of these notions can be applied in the case of India–Pakistan water sharing.

As opposed to the "water wars rationale", the IWT is often equated with "water rationality", which is premised on the understanding that cooperation is a necessity if countries are to ensure long-term access to shared water resources (Alam 2002). In the context of the IWT, it has been contended that despite several factors—including water scarcity, competitive uses of stream flows, territorial contentions, aggressive public statements against the treaty and protracted political rivalry—India and Pakistan chose to negotiate the terms of the treaty over a period of nine years. Although the treaty has been considered sub-optimal, it has also been argued that it was the best possible "cooperative" solution under the circumstances. Scholars have also considered the existence and functioning of the Permanent Indus Commission as a form of "active cooperation" (Zawahiri 2009: 293). The question often raised by water rationality advocates concerns why neither India or Pakistan ever violated the provisions of the treaty so as to safeguard access to shared water even during repeated wars in 1965, 1971 and 1999. Even when several public statements have been issued by leaders on both sides of the border against the treaty itself and water-use by the other side, they have refrained from waging a war over water. In fact, the latest example of World Bank-mediated talks between India and Pakistan—which followed several bellicose public statements by Prime Minister Narendra Modi and other members of the ruling party and declarations regarding proposals to go ahead with a few dam projects in the event of the Uri attacks—shows that the two countries are more inclined to seek negotiated solutions rather than war. However, this has thus far not been translated into an easing of tensions in other areas of shared bilateral interests and engagement.

Yet, the "water wars rationale" continues to be a potent discourse. Both material and ideological concerns have laid at the centre of recurring disputes between India and Pakistan. This is more apparent when the definition of "effective cooperation" is used to analyse the case of the IWT. The treaty, although it offers scope for cooperation—under Article VII—does not contain many cooperative mechanisms beyond data exchange, annual inspections and basic dispute resolution devices. The very fact that the treaty only allocates entire rivers and not partial flows indicates the preoccupation and indeed the need for an independent instead of cooperative water management. Therefore, while India and Pakistan have yet to engage in war over water, this does not mean that such absence of open conflict can be conflated with cooperation. In fact, water management in the subcontinent is heavily shaped by concerns related to "territoriality" and "sovereignty", with control over water and control over territory being positioned along the same continuum. Moreover, the IWT is often seen as an extension of the partition of land in 1947—the IWT simply completing the process by partitioning the waters as well (Sinha 2010: 667). It was equally important for India and Pakistan to have control over resources—in this case, water flows—in order to become completely sovereign; this was more profound in the case of Pakistan for the country claimed historical rights to the waters while leveraging its position as a downstream state. India, on the other hand, wanted to be able to establish its sovereignty, both internally and externally, by not allowing other states to interfere in its water management. Lilienthal's attempt to resolve the problem in a technocratic fashion proved unsuccessful. Despite territorial, ideological and political contentions, one of the reasons why the two governments finally came to an understanding was the influence by the Indus Basin Development Fund (IBDF)—which epitomises the "Cold War development discourse" giving rise to "technocratic internationalism", as argued by Haines (2017). The purely financial explanation of the IWT is, however, replete with strategic overtones as the USA had an interest in South Asian politics, and possibly political stability in the region. Furthermore, both India and Pakistan managed to secure aid through this treaty, in the form of the IBDF. In effect, political and strategic objectives were never compromised in favour of technocratic strategies.

The proposed partition or division of waters was an important requisite for the treaty to be negotiated and put into effect. With hydrological connections cut off as much as possible, the chances of effective cooperation have also become minimal, although one could argue that the two countries still base their behaviour on the treaty and its provisions, especially when a dispute arises. When seen through the prism of peacebuilding, however, this treaty can certainly not be considered a model for advancing cooperation in other arenas or for securing peace and stability between hostile nations. It can at best be considered a mechanism of conflict prevention. Even on the question of water rights, although Pakistan has an upper claim, the rights of the citizenry in the Indian state of Jammu and Kashmir have emerged as a serious concern. Under the treaty, nearly 80% of the waters are allocated to Pakistan, with India having access to only 20%. In recent decades, due to an increasing demand for water, reducing per capita water availability and impacts of climate change on the water resources, Jammu and Kashmir has been affected by severely reduced agricultural productivity and power shortages. Because of the IWT, only 40% of the cultivable land in the state can be irrigated and 10% of the hydro power potential can be harnessed. This has led to criticism of the treaty by India, with several experts and policy-makers accusing the then Prime Minister Jawaharlal Nehru—who signed the IWT—of surrendering to Pakistan and not catering to India's future water requirements, particularly in the states of Jammu and Kashmir, Punjab and Rajasthan. Besides, the treaty lays a strong emphasis on "each party" or "either party", and not on "both parties", thereby leaving very little room for both countries to enter into effective cooperation. Even in the case of cooperation, as specified under Article VII of the treaty, it only refers to engineering works, and no other forms of cooperation that could address environmental and political issues in relation to the basin are contemplated (Chellaney 2011: 96).

Over the years, disagreements over the treaty have increased. These differences regard not only the allocation of waters under different circumstances but also mechanisms of dispute resolution, with Pakistan favouring internationalisation and India preferring to settle disputes bilaterally. In the most recent dispute over the construction of Kishangnaga (inaugurated) and Ratle (construction in progress) projects by India on the River Jhelum and River Chenab, respectively, Pakistan requested the World Bank to empanel the International Court of Arbitration (ICA), arguing that the technical features of these dams are in violation of the IWT as they would interfere with the flow of water into Pakistan. However, India maintains that these are run-of-the-river hydroelectric dams that are well within the parameters outlined in the IWT concerning the western rivers allocated to Pakistan. The Kishanganga project was already approved by the Court of Arbitration in 2013, when Pakistan had taken the issue to it, under the condition that India would maintain environmental flows at all times by releasing a minimum amount of water. However, with differences still not ironed out, India sought the appointment of a Neutral Expert as the objections raised by Pakistan were primarily technical and not linked to any concrete violation of the IWT. There have been two earlier disputes as well—one in the case of the Baglihar Dam on River Chenab, and the other in the case of the Tulbul Navigation Barrage on the River Jhelum, both being India's projects on western rivers. The Baglihar Hydroelectric Power Project was conceived by India in 1992 and the second phase of its construction was completed only in 2008. Pakistan raised objections to certain design parameters of the dam by pointing out that the project could give strategic leverage to India during times of conflict, war or tensions. The matter was taken to the World Bank, which appointed a Neutral Expert—Raymond Lafitte—who then declared his final verdict largely in favour of India, while at the same time upholding some objections raised by Pakistan. This is the only dispute solved as yet, with Pakistan agreeing to abide by the verdict and both sides also eventually resolving differences over the initial filling of the dam in 2010. The Tulbul project, which falls within the parameters for non-consumptive use of navigation, and aimed at enabling trade, tourism and employment, remains under dispute. After Pakistan raised objections to this project (which was envisaged by the Indian side way back in 1987), construction works were suspended until the dispute was resolved (Khadka 2016). While India claims that this project would economically benefit the state of Jammu and Kashmir, Pakistan is wary about the possibility of India using this project to control the flow of water into the River Jhelum (which is one of its objectives), thereby potentially creating drought or flood-like situations in Pakistan. It also fears that India might use the barrage to dry the riverbed in order to expedite movement of its troops to Pakistan during wartime.

Now the question arises as to whether either India or Pakistan are capable of pulling out of the treaty or suspending it partially and/or temporarily—although, under the treaty, this is not an option. Pakistan has repeatedly accused India of violating the provisions of the IWT and threatened that any attempts by India to revoke the treaty would be treated as "an act of war or a hostile act against Pakistan" (Jorgic and Wilkes 2016). Considering that India is the upper riparian, it definitely has the advantage in terms of using the IWT as a bargaining chip against Pakistan when it comes to coercing the latter into accepting certain decisions. At the same time, as mentioned in the draft principles on state obligation drafted by the International Law Commission, an "affected" or "injured" country could resort to countermeasures against another country responsible for causing the injury (Sinha et al. 2012: 748). This assumes significance in the context of the cross-border terrorist activities that Pakistan is accused of perpetrating in Indian Territory. India could revoke the treaty unilaterally as a countermeasure, a move which some experts suggest could be considered well within the parameters of international law. Similarly, the treaty does not take into consideration climate variability, transboundary environmental impact assessment of various activities in the basin, groundwater, environmental flows and so on. On both sides of the border, with population growth, there is an increasing water demand for both irrigation and energy/electricity purposes; and with this growing demand, the pressure on the environment is also mounting—with impacts such as depletion of water resources, pollution (of both groundwater and surface water), and environmental consequences linked to a variety of public projects. With the perpetuation of conflict and the treaty failing to address these concerns, both countries are bound to over-exploit available resources as a mark of their control over the water resources and their ability to deliver to their domestic audiences, much to the detriment of the basin's overall environmental health. Hence, since the treaty does not reflect current realities, especially with respect to environmental factors, could it actually stand the test of time? This is a question that both India and Pakistan need to ponder if they want a long-lasting resolution to the dispute.

5 Impediments to Environmental Peacebuilding

As already stated, the IWT was facilitated by the World Bank's admission that a functionalist approach would not strictly work due to the "historically formed values and beliefs" shaping the relations between India and Pakistan, which were bound to influence the outcome of negotiations over the Indus basin. As Lowi (1993: 67) puts it, the "stronger and less needy upstream state", India, was induced into sitting at the "bargaining table" by the international community, and both sides were urged to separate the issue of river water sharing from larger territorial conflicts—albeit using a non-interdependence approach, as an integrated response was not considered practical. The question that now arises is whether the two are actually separable, as it has been established that water in this context is highly territorialised. Another question that needs to be addressed here concerns whether the argument that an issue such as water scarcity creates interdependencies that necessitate cooperation and mutually beneficial solutions holds in this case. Most importantly, it needs to be verified whether the IWT, the only cooperative arrangement between India and Pakistan as far as river water sharing is concerned, has had spillover effects in other areas of policy-making, in particular those belonging to high politics. The answer is largely "no"—but that does not mean that these assumptions and notions do not have normative value, which could potentially foster opportunities for environmental peacebuilding and translate into better relations between India and Pakistan.

As has been amply explained in the previous sections, the IWT has served as a productive and constructive conflict-prevention mechanism to a great extent, yet its value beyond this function is highly debatable. India and Pakistan have resorted to diplomacy each time a dispute has arisen over the sharing of the River Indus and its tributaries. However, both historical and contemporary analyses show that the IWT is a model of neither "effective cooperation" nor "environmental peacebuilding". The dispute over the Indus basin, which continues to brew, need not and perhaps should not be necessarily be seen in disjunction from the other disputes between India and Pakistan. The conflict regarding Jammu and Kashmir for instance, which is at the heart of the territorial dispute between the two countries, is deeply intertwined with water disputes. In his dissertation, Pervez Musharraf, former President of Pakistan, observed that the Kashmir dispute is mainly premised on the distribution of the waters of the Indus basin between the two countries. He goes on to reiterate that if either of the two disputes were resolved, the other would cease to exist as well (Tripathi, 2011: 70). Although the origin of the Indus lies in the Tibetan Plateau, it flows into Pakistan via Jammu and Kashmir—and this is why there is strong apprehension among the Pakistani authorities that India could divert the waters so as to run Pakistan dry at any point in time. Even though Musharraf's point of view might be overwhelmingly deterministic, it cannot be disassociated from other factors which have worsened India–Pakistan relations over time. It is no surprise that even other disputes like that over the Siachen Glacier is also closely interlinked with the Indus, as the glacier feeds into the Indus river system (via the River Nubra that meets the River Shyok, which in turn eventually meets the River Indus). The strategic importance of Siachen for both India and Pakistan arises not just from the question of control over heights—and, in India's case, for preventing intrusion—but also from its value in the context of the Indus river system.

Political and ideological differences between India and Pakistan have been reflected in their dealings with each other in all sectors, including river water sharing—which is why the IWT was also agreed upon in such a manner that the two countries would not have to jointly manage the river. Not only have the two countries had varying constitutional, political and bureaucratic set-ups, but they have also shared the historical baggage of partition, which was one of the most violent episodes in human history. These factors have invariably pitted the two countries against each other, even on issues of what is considered "low politics"—including the environment, trade, cultural exchange, and so on. There have been many ups and downs in Indo-Pakistan relations (with, many would argue, the latter outnumbering the former). These countries have fought wars, and there have been numerous border killings and other high-profile incidents derailing the peace process (which began in the 2000s under the leaderships of Atal Behari Vajpayee and Musharraf). Further, complicating matters was the addition of a nuclear arsenal to their repository of weapons—the most recent conflict in Kargil in 1999 took place under the nuclear shadow. With both countries going nuclear in 1998, the threat of use of nuclear weapons has also become a part of the discourse on bilateral river water sharing. Moreover, as far as the decision-making system is concerned, the military and intelligence agencies in Pakistan—which are known to be opposed to the development of good relations with India—hold the reins of the country's foreign policy. Pakistan has, therefore and on a few occasions, clarified that it would not hesitate to use its nuclear arsenal against India if the latter chokes the water supply. Although Pakistan does not have an official nuclear doctrine, it has a policy of "first use" in contrast to India's "no first use"; according to lieutenant general (Retired) Khalid Kidwai, the former director general of the Strategic Plans Division of Pakistan, nuclear weapons would be used if the existence of Pakistan as a state was at stake. This was emphasised by President Musharraf before him, but Kidwai further delineated the actions that could put Pakistan's existence at stake, and these include "Indian conquest of Pakistan's territory or military, economic strangling, or domestic destabilisation" (Iqbal 2002). Hence, any attempts to choke the water flow into Pakistan would be treated as a hostile act that could put Pakistan's existence in jeopardy.

One of the biggest impediments to any form of water cooperation in the basin is cross-border terrorism that India accuses Pakistan of perpetrating. In fact, the only time when the IWT came under serious strain was not during a war, but in the aftermath of a terrorist attack in Uri in the state of Jammu and Kashmir, the planning of which could be traced back to Pakistan. Since late 1980s, India has been a victim of cross-border terrorism—allegedly sponsored by Pakistan so as to create instability in India, primarily in Jammu and Kashmir. Despite many terrorist incidents, including high-profile ones such as the Parliament attack in 2001 and the Mumbai attack in 2008, the IWT has so far been left untouched. When the Indian Parliament was attacked by terrorists, masterminded by Jaish-e-Mohammed (JeM) and Lashkar-e-Taiba (LeT), there was enough proof to establish the linkages between these terrorist outfits that operate from Pakistan's territory and the country's Inter-Services Intelligence. However, even at the peak of hostilities, the scheduled meeting of the commissioners of the Permanent Indus Commission took place without any major roadblocks. This was not the case when the 2016 Uri attack was carried out by terrorists of Pakistani origin, which killed 17 Indian soldiers. This was the "deadliest attack on the security forces in Kashmir in two decades" and happened only a few months after the attack on the Pathankot Air Force Station, part of the Western Air Command of the Indian Air Force, which was also linked to the JeM and Pakistan. At this point, the Indian Prime Minister Modi declared that although India would not pull out of the treaty, it would take re-examine and take stock of it and devise mechanisms by which it could utilise the waters of the western rivers that are allocated to Pakistan, which had been underutilised for decades. The Indian administration was under pressure from domestic constituencies to take swift and punitive action against Pakistan after the ghastly terrorist attack on the Indian armed forces; "water" is being a cheaper bargaining chip in comparison with a physical conflict or war, and it was used to put pressure on Pakistan. Ironically, the slogan that Prime Minister Modi used to hit out at Pakistan that "blood and water cannot flow together" was employed by protestors at a march organised by Hafiz Sayeed, an UN-designated terrorist harboured by Pakistan and the mastermind of the Mumbai terrorist attack, back in

2010, when he released a "water declaration", accusing India of perpetrating "water terrorism" against Pakistan, robbing the latter of its share of water by diverting water through projects such as Kishanganga. Farmers, who constituted a large proportion of the protestors carried signs warning, "Water Flows or Blood" (Brulliard 2010). Although both slogans were used as warnings, they served entirely different purposes, influenced by distinct contexts, perceptions and means. Under these circumstances, it is fairly clear that for long-lasting peaceful river water-sharing relations to be achieved, peace in other sectors is essential.

6 Prospects for Environmental Peacebuilding

As already stated, "environmental peacebuilding" as a normative concept does have implications for the subcontinent. Therefore, the best way to conclude this chapter would be to elaborate upon prospects for environmental peacebuilding between India and Pakistan. It is largely acknowledged by everyone from the scientific, legal and policy communities that the IWT needs to be either reviewed or replaced by another treaty that addresses current realities of environmental degradation, climate change, domestic requirements and geopolitical dynamics. The IWT, as a short-term measure, served its purpose of avoiding conflict and addressing threats of various types. With changing environment and demographics, the treaty may not be good enough to cater to growing water demands or find solutions to worsening water stresses in the basin. These changes are putting immense pressure not only on water security but also food and energy security, particularly in Pakistan. The IWT was negotiated on the premise that there was sufficient water for both parties, yet this is not the case anymore. A satisfactory utilisation of the waters by both parties, as envisioned in the IWT, entails more efficient and sustainable uses, for which strict division of waters would prove counterproductive. Under the current frame, both countries maintain overarching sovereignty over their share of resources and over-exploit them without wider considerations regarding the ecosystem as a whole—and this state of affairs needs to be replaced by joint management initiatives. The treaty also cannot be completely disassociated from internal politics in either India or Pakistan. For instance, in Pakistan there exists rivalry among the provinces with respect to the use of the basin's waters—as provinces such as Sindh and Balochistan accuse Punjab of denying them proper water rights. Since the irrigation systems were developed in Punjab—the bread basket of Pakistan—a large proportion of the water of the Indus and its tributaries is used in this province, especially through the construction of dams. This has led to rivalries between Punjab and other provinces such as Sindh and even Khyber Pakhtunkhwa (Ranjan 2015). The Pakistan Apportionment Accord, signed in 1991 between Punjab, Baluchistan, Sindh and Khyber Pakhtunkhwa, has failed to resolve differences over allocation and supply of water among these provinces (Anwar and Bhatti 2018). In India too, Punjab, Haryana and Rajasthan have not been able to find a sustainable solution to their disputes over the Ravi and the Beas, two tributaries of the River Indus. In 2004, Punjab unilaterally passed an act—the Punjab

Termination of Agreement Act—by which it abrogated its pacts concerning the sharing of the Ravi and the Beas with Rajasthan and Haryana, the latter state having been carved out of the state of Punjab in 1981. Although India's apex court, the Supreme Court, called this decision unlawful, it needs to be taken into consideration that despite Rajasthan and Haryana not being riparians of the tributaries of the Indus, they possess water rights in relation to these rivers. When the IWT was negotiated, Haryana did not exist (since it was a part of Punjab), and so concerns related to non-riparian rights and how they can be integrated within the IWT needs to be dealt with, especially since the agricultural landscape of Haryana has changed dramatically over the years. Moreover, due to flawed water and electricity policies, the conflict between agricultural and industrial users over limited resources is growing; in addition, water tables are declining at a rapid rate. However, for these internal tussles as well, the IWT is blamed. Therefore, more than physical water scarcity and the question of supply versus demand, it is the politics of distribution and access that needs to be looked into; it is the socio-economic and material infrastructure that requires to be rectified and/or refurbished. In effect, water scarcity in the region that depends on the basin has increasingly raised questions on the utility and relevance of the IWT in contemporary times.

The case of groundwater depletion, which is perhaps the most pressing issue in South Asia, more specifically in India and Pakistan, is not addressed by the IWT. When it comes to surface water, the treaty clearly lays out the rules, yet in terms of groundwater management there is a massive policy vacuum. The Indus Basin is one of the most environmental stressed basins in the world. According to a NASA study, it is the second most stressed basin on the planet—falling by 4–6 mm per year. While everyone talks about conflict over surface water, it is groundwater that could trigger a conflict between the two countries as declining water tables are most likely to have cross-border impacts. India and Pakistan are the first and fourth largest users of groundwater in the world, respectively. 60% of India's irrigated area is groundwater-dominated; groundwater utilisation is extremely high in the states of Punjab, Haryana and Rajasthan—two of which share a border with Pakistan. If one takes the case of Pakistan, in the Punjab province (which produces 90% of the country's food) groundwater caters to more than 40% of the total crop water requirements (Shah 2007: 7). Furthermore, in both countries, there are no clear definitions of groundwater entitlements and usually groundwater access is determined strictly by land ownership—which leads to over-extraction and exploitation of groundwater well beyond recharge capacity. Moreover, many crops grown in the Indus Basin such as wheat, rice, cotton and sugarcane are water-intensive and entail flood irrigation—leading to further groundwater depletion, which is aided by a subsidised electricity supply (used for groundwater pumping). Groundwater depletion is causing a slew of environmental problems such as soil salinisation and land subsidence, which in turn affect agricultural productivity and increase the costs of cultivation in general (Watto and Mugera 2016). In addition, a 2013 report by the International Union for Conservation of Nature (Pakistan) observes that India's dam projects could induce groundwater recharge in its territory but would restrict the flow of surface water into Pakistan owing to "seepage losses in lakes and reservoirs".

The level of uncertainty in the Indus Basin is multiplied by various other factors, the most important of which is probably environmental and climate change. The impacts of climate change could manifest themselves in many ways, but mainly by affecting the "timing and magnitude of annual precipitation and the rate and significance of deglaciation in the Himalayas and sub-Himalayan mountain ranges". The Indus Basin's dependence on both monsoon and glacial meltwater is high—and with climate change; these two phenomena are expected to become extremely uncertain, adding to the stress of the basin, particularly in the long term. For instance, although the degree of contribution of glacial meltwater to the rivers' flow is contested, it is largely acknowledged by the scientific community that reduced meltwater is expected to have adverse impacts on river runoff in the Indus Basin (Miller, Immerzeel and Rees 2012: 463–464). According to an estimate, "Based on current projections, the Indus River system is expected to fall below 2000 flow levels between 2030 and 2050. The drop-off is estimated to be most serious between 2030 and 2040, with a new equilibrium flow of 20% below that of 2000 reached after 2060". This would put a significant pressure on the per capita availability of water in the region, which is already dwindling at a fast rate. The per capita availability of water is on a steep decline in both countries—in Pakistan, it is less than 1000 m^3 (Ebrahim 2018); in India, the 2011 census showed the per capita availability of water as 1545 m^3 (https://pib.gov.in/newsite/printrelease.aspx?relid=119797).

Even if one takes the case of the River Nubra, which is fed by meltwater of the Siachen glacier, we see that it is affected not only by climate change but also by the sheer presence of both countries' troops and associated pollution. In fact, Pakistan has occasionally used the argument that Indian troops in Siachen are "posing a serious threat to Pakistan's environment by damaging Siachen's virgin snow" and "damaging one of the largest sources of water to Pakistan on a regular basis" (Philip 2013). The Indus Basin is one of the most polluted in the world, known for its dubious distinction as one of the ten rivers in the world that contributes more than 90% of the plastic entering the world's oceans, according to a report published by the Helmholtz Centre for Environmental Research in 2017. On both sides of the border, agricultural, industrial and urban discharge into the basin's rivers has put public health and ecosystems at stake, endangering the health of the basin itself (Pappas 2011). In addition, the basin is highly disaster-prone and has seen several floods and droughts in the past decades. Yet not much attention has been paid to addressing the challenges posed by disasters through risk reduction and mitigation. When the IWT was signed, climate change was not a factor; however, today drastic and uncertain changes in the environment and climate make it disastrous to continue with a treaty that does not even take them into account.

It is important to highlight the geopolitical aspects of the Indus basin, which are central to India–Pakistan river water-sharing relations. Although India is the upper riparian, the headwaters of the Indus are controlled by China; this is a factor that cannot be overlooked while exploring prospects for cooperation through integrated basin management. Any actions by China in the basin, especially the construction of dams, could have serious repercussions for both India and Pakistan. In addition, China has also agreed to finance dams in the disputed territory of Gilgit-Baltistan as a part of the China–Pakistan Economic Corridor (CPEC). For example, the proposed Diamer-Bhasha Dam on the River Indus, which was denied funding by the World Bank and Asian Development Bank due to India's objections as the region is claimed by India as its territory, has been approved for Chinese funding (Neelakantan 2017). Not only are these dams detrimental to Pakistan's own economic interests as they could diminish the flow of silt that is a necessary for agriculture downstream and the flow of water in the non-monsoon season (Gupta 2017), also they complicate the geopolitics of the region further by reinforcing the links between basin and sovereignty, thereby reducing chances of cooperation. There does not seem to be enough discussion on dams on the River Indus and its tributaries being constructed in Chinese territory such as the medium scale one near Demchok (Ladakh), which could restrict water flows considerably and affect energy and irrigation requirements downstream in both Pakistan and India. For that matter, the mega hydroelectric projects proposed to be constructed in Gilgit-Baltistan would face water shortages as well (Sering 2010). Since China has not signed any agreement with either India or Pakistan, it is rather difficult to envisage "effective cooperation" in the Indus Basin in the near future. Therefore, any prospects for cooperation are, to some extent, contingent on bringing other countries such as China and Afghanistan on board. The geopolitical aspects are not confined to the Indus basin alone; if India chooses to restrict the flow of the Indus and its tributaries to Pakistan, this could set a precedent for a country like China to do the same—which could eventually backlash against India. In fact, just days after India warned Pakistan that it would claim its rightful share under the IWT, China announced that it would block an important tributary of the River Brahmaputra— the River Xiabuqu—in order to construct the "most expensive hydropower project" to date. Even though the project might not have significant effect on downstream flows, it is bound to raise apprehension among the Indian leadership regarding future Chinese projects on River Brahmaputra and its tributaries (Krishnan 2016).

There is enough scope for cooperation between India and Pakistan. From integrated river basin management, which is a holistic water security proposal, to a sub-regional hydro-economy, which serves material interests of both countries, there are ways in which environmental peacebuilding can be set in motion. An integrated river basin management mechanism that takes into consideration issues such as accountability, transparency, engagement and sound design based on technical, scientific and other information, advocacy and institutionalisation, and so on, might be a far cry in the subcontinent. However, scholars have proposed—for instance—the establishment of a joint and independent river basin organisation for this purpose by renegotiating the treaty (Swain 2009). The steps towards this goal could be built by taking initiatives such as setting up a strategic knowledge platform for climate change that could integrate with national programmes on climate change such as India's Mission on Strategic Knowledge for Climate Change (Ali and Zia 2017: 131). Since there is a lack of joint efforts to study the impacts of climate and environmental change in the basin and the region in general—except through institutions like the International Centre for Integrated Mountain Development (ICIMOD) targeted at the Himalayan region and the South Asian Association for Regional Cooperation (SAARC) which has so far not fetched any fruitful results—it is of utmost importance for the two countries to engage in collaborative and/or joint research ventures to design better response mechanisms—both policy and technical such as weather forecasting and early warning systems. The SAARC has also not been able to catalyse regional climate change mitigation and adaptation programmes in the region. Considering that India and Pakistan are extremely vulnerable to climate change and that the basin is highly climate-stressed, they need to channel their efforts into discussing potential joint mechanisms that can tackle the issue regionally (as much as nationally). Without dealing with climate change at the basin-level, there cannot be a comprehensive solution to the challenges that climate change poses to the entire region. At the same time, it is of essence for the two countries to look beyond the IWT and engage other stakeholders in the dialogue related to river water sharing; unless there is greater awareness of water-related problems on both sides of the border, the dynamics can never change. In fact, if the primary stakeholders are brought to the table, chances are that the "real" issues are focussed upon more frequently, thereby also enhancing the prospects for cooperation—and, in the long run, peacebuilding.

India has been cooperating with Bhutan and Nepal for decades now in terms of harnessing hydro-potential of these Himalayan nations—albeit one could always debate the power dynamics and outcomes in question. The sub-regional framework comprising Bangladesh, Bhutan, India and Nepal (BBIN) has been emphasising the need for improving connectivity between the four nations, especially in fields like energy (mainly hydropower). This is the reason for the recent announcement of approval of Indian land for establishing transmission infrastructure for the Bangladesh–Nepal electricity trade (Sarkar 2018). One could argue that since both India and Pakistan (especially the latter) are facing power shortages, it is prudent for them to cooperate over hydroelectricity projects rather than clashing over them. It could be contended that the material or infrastructure-led cooperation could spell disaster for the basin as both existing and proposed projects could have dire ecological consequences; as already described in the chapter, this has tended to inflame tensions between the two countries as well as between provinces or states. These projects need to be designed in such a way that both ecological and economic gains are balanced out. Similarly, the groundwater recharge situation calls for recharge and management initiatives that ensure sustainable groundwater extraction. This would entail an overhaul of agricultural practices on both sides of the border; hence, they need to be tied to both intrastate policies and interstate talks concerning water. For this, information sharing on recharge and discharge rates, as well as other scientific and technical data, holds the key to integrated basin management.

7   Conclusion

India–Pakistan relations have been historically unstable and are expected to continue so in the near future. In such a scenario, the chances of water being used as a leverage or held as hostage in hostile geopolitical dynamics are high. One could argue that the conflict could lead to a more sustainable use of water, but in the subcontinent it has been seen that the conflict has instead driven both countries to over-exploit resources, aggravating water insecurity and ecological destruction. The future trajectory seems to be the same—with the recent announcements by both governments regarding the construction of more infrastructure. Even though some newspaper reports (mainly in Pakistan) have indicated that India plans to build over 10 dams, this cannot be corroborated and can at best be treated as propaganda intended to build Pakistan's case against India internationally. At the same time, India's green light to a hydroelectric project on the River Chenab in Jammu and Kashmir and its proposal to renew the Tulbul Navigation Project are clear signals of India's intent. Nevertheless, it needs to be noted that the geographical realities dictate the future of bilateral relations in the basin. For example, the biggest impediment for India's aspiration to divert the waters of the Indus and its tributaries is geography—as the terrain, along with the country's fragile storage and irrigation systems, makes diversion extremely difficult due to the staggering economic costs involved in building such infrastructure in the Himalayan region. Any attempts by India to restrict the flow of water beyond a point could lead to flooding in Indian states, as the country does not have adequate storage infrastructure (Parvaiz 2016) (Guruswamy 2016). With these new dam structures, India will be in a better position to control water flows, but that is a long way away. Pakistan, on the other hand, has to come to terms with these geographical realities, so central to river water-sharing dynamics.

With these geographical factors playing a significant role in ensuring that there is very little scope for water flows to be restricted at least in the short and medium terms, the two countries have an opportunity to shift attention to effective water cooperation. It is amply clear that the Indus basin is extremely stressed both environmentally and climatically, and that without proactive steps being taken by the riparian nations to protect the basin jointly, disaster could ensue. The IWT is not a sufficient mechanism to address water insecurity issues in the region as it only divides the rivers as a whole and does not go into the nitty-gritty of water management. There is no dearth of opportunities for water cooperation which could translate into scientific and cultural exchanges or economic cooperation, as they are interconnected to a large extent. In the long term, the economic, social and cultural spin-offs from water cooperation are capable of engendering peace initiatives by strengthening peace talks, which so far have failed over and over again. However, ideological and territorial faultiness are too deep for any far-reaching steps to take off. In such a situation, the best possible option could be to retain the IWT and work around Article VII—which leaves some room for cooperation but does not provide specific details, thereby allowing the two countries to set goals realistically. In short, beyond the normative dimensions of environmental peacebuilding, it could be argued that unless there is political and territorial stability (particularly in terms of bilateral relations), environmental and climatic initiatives are either a non-starter, or at best could have only short-term success.

References

About Environmental Peacebuilding (2018) Environmental peacebuilding. Available at: https://environmentalpeacebuilding.org/about/. Accessed 25 Nov 2018

Aggestam K (2015) Desecuritisation of water and the technocratic turn in peacebuilding. Int Environ Agreem: Polit Law Econ 15(3):327–340

Aggestam K, Sundell-Eklund A (2014) Situating water in peacebuilding: revisiting the Middle East peace process. Water Int 39(1):10–22

Alam UZ (2002) Questioning the water wars rationale: a case study of the Indus Waters Treaty. Geogr J 168(4):341–353

Ali SH, Vladich HV (2016) Environmental diplomacy. In: Constantinou CM, Kerr P, Sharp Paul (eds) The SAGE handbook of diplomacy. Sage, New York

Ali SH, Zia A (2017) Transboundary data sharing and resilience scenarios: harnessing the role of regional organizations for environmental security. In: Adeel Z, Wirsing RG (eds) Imagining Industan: overcoming water insecurity in the Indus Basin. Springer, Switzerland

Anwar AA, Bhatti MT (2018) Pakistan's water apportionment accord of 1991: 25 years and beyond. J Water Resour Plan Manag 144(1):05017015

Axelrod R, Keohane R (1986) Achieving cooperation under anarchy. In: Oye K (ed) Cooperation under anarchy. Princeton University Press, Princeton

Baechler G (1998) Why environmental transformation causes violence: a synthesis. Environ Change Secur Proj Rep 4:24–44

Barnett J (2000) Destabilizing the environment-conflict thesis. Rev Int Stud 26(2):271–288

Bhatnagar M (2014) Reconsidering the Indus Waters Treaty. Tulane Environ Law J 22(2):271–313

Brulliard K (2010) Rhetoric grows heated in water dispute between India, Pakistan. Washington Post, 28 May. Available at: http://www.washingtonpost.com/wp-dyn/content/article/2010/05/27/AR2010052705393.html??noredirect=on. Accessed 23 Nov 2018

Chellaney B (2011) Water: Asia's new battleground. Georgetown University Press, Washington, DC

Claude IL Jr (1956) Swords into ploughshares. Random House, New York

Conca K, Dabelko GD (2002) Environmental peacemaking. Woodrow Wilson Center Press, Washington, DC

Conca K, Carius A, Dabelko GD (2005) Building peace through environmental cooperation. In Worldwatch Institute, State of the World (2005) Redefining Global Security. Norton, New York

Deudney D (1990) The case against linking environmental degradation and national security. Millennium 19(3):461–476

Dinar S (2011) Beyond resource wars: scarcity, environmental degradation, and international cooperation. MIT Press, Cambridge

Ebrahim ZT (2018) Is Pakistan running out of fresh water? Dawn, 30 March. Available at: https://www.dawn.com/news/1398499. Accessed 28 Nov 2018

Gleditsch NP (1998) Armed conflict and the environment: a critique of the literature. J Peace Res 35(3):381–400

Gleick PH (1991) Environment and security: the clear connections. Bull At Sci 47(3):16–21

Gupta J (2017) Indus cascade a Himalayan Blunder. The Third Pole, 22 May. Available at: https://www.thethirdpole.net/en/2017/05/22/indus-cascade-a-himalayan-blunder/. Accessed 20 Nov 2018

Guruswamy M (2016) Indus waters treaty: turning off the taps to Pakistan. DailyO, 29 Sept. Available at: https://www.dailyo.in/politics/indus-waters-treaty-uri-attack-pakistan-kashmir-valley-narendra-modi-nawaz-sharif/story/1/13162.html. Accessed 20 Nov 2018

Haines D (2017) Rivers divided: Indus Basin waters in the making of India and Pakistan. Oxford University Press, Oxford

Henriette L (2012) Regimes in Southeast Asia: an analysis of environmental cooperation. Springer, Berlin

Hensel PR, Mitchell SM, Sowers TE II (2006) Conflict management of riparian disputes. Polit Geogr 25(4):383–411

Homer-Dixon TF (1999) Environment, scarcity, and violence. Princeton University Press, New Jersey

Huntjens P, Yasuda Y, Swain A, de Man R, Magsig B-O, Islam S (2016) The multi-track water diplomacy framework: a legal and political economy analysis for advancing cooperation over shared waters. The Hague Institute for Global Justice, The Hague

Ide T, Detges A (2018) International water cooperation and environmental peacemaking. Glob Environ Polit 18(4):63–84

International Union for Conservation of Nature and Natural Resources (2013) Beyond Indus water treaty: water cooperation for managing groundwater environments—policy issues and options. IUCN Pakistan, Islamabad

Iqbal N (2002) Economic threat may push Pakistan to go nuclear. Asia Times, 6 Feb. Available at: http://www.atimes.com/ind-pak/DB06Df02.html. Accessed 23 Nov 2018

Islam S, Repella AC (2015) Water diplomacy: a negotiated approach to manage complex water problems. J Contemp Water Res Educ 155(1):1–10

Jervis R (1978) Cooperation under the security dilemma. World Polit 30:167–214

Jorgic D, Wilkes T (2016) Pakistan warns of 'water war' with India if Indus water treaty violated. Livemint, 27 September. Available at: https://www.livemint.com/Politics/QUAAXqgoReMYWhPjoPvltI/Pakistan-warns-of-water-war-with-India-if-Indus-water-trea.html. Accessed 23 Nov 2018

Kaplan RD (1994) The coming anarchy. The Atlantic, February

Khadka NS (2016) Are India and Pakistan set for water wars? BBC, 22 Dec. Available at: https://www.bbc.com/news/world-asia-37521897. Accessed 23 Nov 2018

Krishnan A (2016) China Blocks Tributary of Brahmaputra in Tibet to Build Dam. India Today, 2 Oct. Available at: https://www.indiatoday.in/mail-today/story/most-expensive-dam-brahmaputra-tributary-china-344357-2016-10-02. Accessed 20 Nov 2018

Kyrou CN (2007) Peace ecology: an emerging paradigm in peace studies. Int J Peace Stud 12(1):73–92

Lowi MR (1993) Water and power: the politics of a scarce resource in the Jordan River Basin. Cambridge University Press, Cambridge

Maas A, Carius A, Wittich A (2013) Environmental cooperation as a tool for peace-building. In: Floyd Rita, Matthew RA (eds) Environmental security: approaches and issues. Routledge, London

Matthews JT (1989) Redefining security. Foreign Aff 68(2):162–177

Miller JD, Immerzeel WW, Rees G (2012) Climate change impacts on glacier hydrology and river discharge in the Hindu Kush–Himalayas. Mt Res Dev 32(4):461–467

Miner M, Patankar G, Gamkhar S, Eaton DJ (2009) Water sharing between India and Pakistan: a critical evaluation of the Indus Water Treaty. Water Int 34(2):204–216

Myers N (1996) Ultimate security: the environmental basis of political stability. Island Press, New York

Neelakantan S (2017) China to Build Dam in Pakistan that World Bank, ADB Refuse to Fund. The Times of India, 20 June. Available at: https://timesofindia.indiatimes.com/world/pakistan/china-will-make-dam-project-india-objects-to-part-of-its-cpec-project-in-pakistan-says-islamabad/articleshow/59229530.cms. Accessed 20 Nov 2018

Pappas G (2011) Pakistan and water: new pressures on global security and human health 101(5):786–788

Parvaiz A (2016) India simply cannot afford to scrap the Indus Water Treaty with Pakistan. Quartz, 26 Sept. Available at: https://qz.com/india/790885/narendra-modi-led-india-simply-cannot-afford-to-scrap-the-indus-water-treaty-with-pakistan/. Accessed 20 Nov 2018

Petersen-Perlman JD, Veilleux JC, Wolf AT (2017) International water conflict and cooperation: challenges and opportunities. Water Int 42(2):105–120

Philip SA (2013) Pakistan asks India to withdraw army from Siachen Glacier. Livemint, 4 Dec. Available at: https://www.livemint.com/Politics/xpxaOJNcQDbyjfYbArntoI/Pakistan-asks-India-to-withdraw-army-from-Siachen-glacier.html. Accessed 23 Nov 2018

PTI (2017) SC Rejects Plea to Declare Indus Waters Treaty as Illegal. The Hindu, 10 Apr. Available at: https://www.thehindu.com/news/national/sc-rejects-pil-to-declare-indus-waters-treaty-unconstitutional/article17903227.ece. Accessed 20 Nov 2018

Roic K, Garrick D, Mansoor Q (2017) The Ebb and flow of water conflicts: a case study of India and Pakistan. In: Adeel Z, Wirsing RG (eds) Imagining Industan: overcoming water insecurity in the Indus Basin. Springer, Switzerland

Ranjan A (2015) Water disputes between Punjab and Sindh: a challenge to Pakistan. New Water Policy Pract J 1(2):46–58

Romm JH (1993) Defining national security: the nonmilitary aspects. Council on Foreign Relations, New York

Sadoff CW, Grey D (2002) Beyond the river: the benefits of cooperation on international rivers. Water Policy 4:389–403

Sarkar D (2018) India to Catalyze Bangladesh-Nepal Bilateral Power Tie Up. The Economic Times, 20 Feb. Available at: https://economictimes.indiatimes.com/industry/energy/power/india-to-catalyze-bangladesh-nepal-bilateral-power-tie-up/articleshow/62998747.cms. Accessed 20 Nov 2018

Sering SH (2010) China builds Dam on Indus near Ladakh. J Def Stud 4(2):136–139

Shah T (2007) The groundwater economy of South Asia: an assessment of size, significance and socio-ecological impacts. In: Giordano M, Villholth KG (eds) The agricultural groundwater revolution: opportunities and threats to development. CABI, Wallingford

Sinha UK (2010) 50 years of the Indus water treaty: an evaluation. Strat Anal 34(5):667–670

Sinha UK, Gupta A, Behuria A (2012) Will the Indus water treaty survive? Strat Anal 36(5):735–752

Sjöstedt G (2009) Resolving ecological conflicts: typical and special circumstances. In: Bercovitch J, Kremenyuk V, William Zartman I (eds) The SAGE handbook of conflict resolution. Sage, London

Swain A (2009) The Indus II and Siachen peace park: pushing the India-Pakistan peace process forward. Round Table 98(404):569–582

Tarlock D (2015) Promoting effective water management cooperation among riparian nations. Global Water Partnership, Stockholm

The Indus Waters Treaty (1960) Available at: https://siteresources.worldbank.org/INTSOUTHASIA/Resources/223497-1105737253588/IndusWatersTreaty1960.pdf. Accessed 20 Nov 2018

The Punjab Termination of Agreement Act (2004) Available at: https://punjabxp.com/punjab-termination-agreement-act2004/. Accessed 20 Nov 2018

Tripathi NK (2011) Scarcity dilemma as security dilemma: geopolitics of water governance in South Asia. Econ Polit Wkly 46(7):67–72

Ullman RH (1983) Redefining security. Int Secur 8(1):129–153

United Nations Environment Programme (2015) Addressing the role of natural resources in conflict and peacebuilding: a summary of progress from UNEP's environmental cooperation for peacebuilding programme. UNEP, Nairobi

United Nations Interagency Framework Team for Preventive Action (2012) Toolkit and guidance for preventing and managing land and natural resources conflicts: renewable resources and conflict, Nairobi. Available at: http://www.un.org/en/land-natural-resources-conflict/pdfs/GN_Renew.pdf. Accessed 28 Nov 2018

Watto MA, Mugera AW (2016) Groundwater depletion in the Indus Plains of Pakistan: imperatives, repercussions and management issues. Int J River Basin Manag 14(4):447–458

Weiss JN (2003) Trajectories towards peace: mediator sequencing strategies in intractable communal conflicts. Negot J 19(2):109–115

Zawahiri N (2009) Third party mediation of international disputes: lessons from the Indus River. Int Negot 14(2):281–310

Zeitoun M, Mirumachi N (2008) Transboundary water interaction I: reconsidering conflict and cooperation. Int Environ Agreem 8:297–316

Zeitoun M, Warner J (2006) Hydro-hegemony: a framework for analysis of transboundary water conflicts. Water Policy 8:435–460

Chapter 12
Rainfall Deficiency and Water Conservation Mechanisms in Western Rajasthan

Ujjwal Dadhich and Abdul Shaban

1 Introduction

Western Rajasthan is a water deficit region. The majority of the landscape is desert-like. Low rainfall, frequent droughts, extreme temperature differences, high-speed winds and the absence of perennial water resources make living condition harsh (The Rajputana Gazetteer 1879; Tod 1902, 1920; Hora 1952; Khosla 1952; Directorate District Gazetteers 1995; Dhir and Singhvi 2012; ISRO 2016; Kar 2017). However, communities have tried to adapt to the harsh living conditions by devising various innovative methods so as to manage the environment. Despite chronic water shortages and consequent limitations to agriculture and related activities, the area has been occupied for centuries and even witnessed several progressive kingdoms (The Rajputana Gazetteer 1879; Tod 1920; Mishra 1993, 1994). Oral and folklore traditions (Bharara 1980; Bharara and Seeland 1994), written records and more importantly the structural remains (both abandoned and still functioning) (Mishra 1993, 1994; Levi and Mishori 2015) help us understand the water management techniques devised. These sources point out that successful management in such conditions was primarily due to a lifestyle based on symbiosis with nature. That is, communities directed their everyday habits and livelihood activities so as to take natural constraints into account. These communities adopted a number of strategies and constructed structures intended to store scarce water for longer, recycled and

U. Dadhich (✉) · A. Shaban
School of Development Studies, Tata Institute of Social Sciences, Deonar, Mumbai, India
e-mail: dadhichujjwal007@gmail.com

A. Shaban
e-mail: shaban@tiss.edu

© Springer Nature Switzerland AG 2020
S. Bandyopadhyay et al. (eds.), *Water Management in South Asia*,
Contemporary South Asian Studies, https://doi.org/10.1007/978-3-030-35237-0_12

multiple uses. These structures ensured water security and sustainable water supply for drinking, livestock and agricultural purposes. This paper discusses such important structures—naadi, khadin, taanka, baori and beri—which evolved historically in the region through a society–environmental interface. The construction of such structures displays technical knowledge gained from accumulated experience. Taken together, these structures constitute participative and decentralized water governance arrangements. One may consider such structures historical and outdated. Yet they are still functional across different regions. Some of these structures have also undergone some improvements. In fact, today a number of villages, civil society organizations and individuals have resorted to these structures so as to mitigate increasing water crises. The paper discusses these aspects and suggests a combined use of traditional and modern technology for efficient water management in the regions.

2 Geographical Conditions in Rajasthan

Suraj ki hai yeh dharti, Anant sagar tak hai yaha ret, Saari nadiya sookh gayi, ab bache hai keval avshesh, Phir bhi yaha base Rajwaade aur baad mein aaye Angrez Banaye bade shahar aur seeche gaye khet Kaha se aaya paani aur kaise geeli hui ret?

Local saying

This is the land of the Sun, there is an ocean of sand, all rivers have dried up, leaving but traces. Still kingdoms were there; later British Huge cities were constructed, much farmland was irrigated. Where did the water come and how was the thirsty sand quenched?

Translated version

The above couplets from a local saying highlight the difficult environmental conditions as well as human survival instincts and innovative spirit.

As mentioned above, living in western Rajasthan is not easy. The desert is wide and associated climatic conditions are harsh. Local folklores cite these conditions by personifying the Sun as king of this desert. The will of the Sun defines living conditions—i.e., temperature, cloud formation, wind and rainfall. Those who challenge this state of affairs are doomed and will perish. This has been the case with local rivers, which have now dried up so that there is water deficiency everywhere. One therefore cannot challenge the Sun but has to live under its mercy. One has to learn from the Sun and from these climatic conditions, not try to challenge them.

Local sayings are just reflections of what communities see and experience. The above description therefore provides a local perspective of existence in western Rajasthan. These views are further validated by both quantitative figures and scientific studies. For example, nearly 62% of Rajasthan's administrative area[1] (approximately 205,583 km^2) is now desertified (ISRO 2016). This covers the entire Marwar[2] and parts of the Shekhawati regions. It includes 13 districts—particularly Jaisalmer, Bikaner, Jodhpur, Nagaur, Barmer, Jalore, Churu, Jhunjhunu, Sikar, and parts of Pali, Sri Ganganagar and Sirohi. This is a huge area. In fact, the desert was once a transgressional sea. Glacial winds from the North and North-West (Central Asia) used to bring heavy rains here. The rise of the Aravalli Range and later the Himalayas in the Miocene period however obstructed these winds and rainfall, and the area became a desert (Banerji 1952; Krishnan 1952; Ahmed 1969; Singh 1977; Dhir and Singhvi 2012).

This desert has become entirely dependent on the south-west monsoon from the Arabian Sea for even scanty rainfall. However, these winds usually just blow through the region without delivering much water. This can be attributed to two reasons. First, the desert area is a low-lying region, and there is no appreciable relief present in the wind course which could to obstruct it and lead to orographic rainfall. As the Aravalli Range is parallel to wind movements, it does not offer much obstruction.

It would be interesting here to examine how nature works inside the desert. How do different natural elements interact and build networks that together shape the desert landscape? Sand, wind, temperature and landforms will now be of more concern here. We attempt to highlight their function through local sayings. Some of the local sayings personify these elements as soldiers of the Sun's army. These soldiers are always awake; they guard the desert and support life in their own ways. A local couplet provides this understanding of the desert:

> Pawan chalti jaave, kabhi dhora ne jama, kabhi kaat ti jaaye Suraj ko taap ye milaap karave, chaaro aur tilla banta jaave.
>
> Respondent
>
> The wind blows, it sometimes deposits and sometimes erodes the sand. The Sun's heat causes the interaction; all around dunes begin to stand.
>
> Translated version

Sand and wind are the observable features of the desert. They shape the desert. The soil in the region is largely sandy-loamy in nature (Tamhane 1952; Roy and Sen 1968; Shyampura and Sehgal 1995). These dry sand particles are carried away by

---

[1]Rajasthan, spread over an area of nearly 342,239 square kilometre (km), is presently the largest Indian state. It is located in the North-West and occupies nearly 10.4% of the country's total area (ISRO 2016). The state shares an international border of nearly 1070 km with Pakistan in the west and has internal borders with the neighbouring states of Punjab (in the North), Haryana and Uttar Pradesh (in the North-East), Madhya Pradesh (in the South-East) and Gujarat (in the South-West).

[2]Topographical, cultural and social/clan differences led to a division of Rajasthan into nine linguistic regions (Tod 1902; Office of The Registrar General & Census Commissioner 2011). These includes Ajmer Merwara (in the North Central region), Dhundhar (comprising the present-day districts of Jaipur, Dausa, Madhopar and Tonk in the Central East), Gorwar (Sirohi, Jalore and Pali districts in the South-West), Hadoti (Bundi, Baran and Kota districts in the South-East), Marwar (Jodhpur, Jaisalmer, Bikaner, Barmer and parts of the Pali district in western Rajasthan), Mewar (Bhilwara, Chittorgarh and Udaipur in the South Central region), Mewat (Alwar, Bharatpur and Dholpur district in the Eastern region), Shekhawati (Jhunjhunu and Sikar districts in the North-Eastern region) and Vaggad (Banswara and Dungarpur district in the South).

the wind to different places. Sand dunes here can be as high as 500 m and may have a thickness of up to 100 m. As we go deeper into the desert, dune density increases. Locals call this area thali. This area presents a spectacular view, especially at Sunrise and Sunset. One can enjoy the shadows which dunes cast on each other and also the panoramic, golden and colourful combination of light and light reflections. The entire scenic beauty seems to be a beautiful painting painted by the Earth itself and its different elements.

In western Rajasthan, maximum temperature during the hot summer days can reach up to 50 °C. The hot sand however cannot retain the heat and loses it rapidly by night. This causes the temperature to drop—and it can reach up to a minimum of 0–2 °C during the winter. Large temperature fluctuations cause high wind velocities, sometimes around 50–75 km/h. This leads to large-scale Aeolian activities during the months between March and July (Kaul 1996). High-impact erosion and deposition occur during this period. As a result, existing sand dunes are eroded, new ones are formed, and some dunes become larger due to particle accumulation. All this happens at a great frequency. Dunes now appear to be shifting, and it is really difficult to make sense of their original locations. Because of these features, the Thar Desert is also known as the land of shifting dunes. These shifting dunes are locally known as dhrians (The Rajputana Gazetteer 1879). Such large-scale Aeolian activities are harmful to the local agriculture and also to livestock. It wears off the upper fertile soil and also uproots many saplings and crops. But even in such difficult situations, locals have their own adapting and mitigating mechanisms. A local resident remarks,

> …Garmiyon mein jaise Suraj chadhta hai, yaha ret ka taap bhadta hai Loo ke thapede chalte hai, kabhi bawandar bhi uthta hai Dhore bhi idhar udhar chalte hai, ye mushkil ki ghadi hoti hai lekin mujhe ye bhi Prakriti ka utsav lagta hai…
>
> A local saying

> …As the Sun rises during summer, the temperature increases as the day Loo (hot winds) blows, and sometimes sandstorms also come. Sand dunes start moving here and there, for us it is a difficult condition. But to me it is also a festival of nature…
>
> Translated Version

In addition to the dunes, there are also other landforms in the desert landscape. Prominent among these are old aggraded alluvial plains and salt lakes (Dhir and Singhvi 2012). The alluvial plains can be observed in the north and west of Jodhpur, parts of Nagaur, Churu and even in the arid regions of Jaisalmer, Bikaner and Jalore. Such an extensive presence of alluvial plains is largely due to the river drainage systems of Ghaggar, Luni and Saraswati rivers. Ghaggar and Luni are the only seasonal rivers in the region. They were in fact tributaries of the River Saraswati which has now vanished[3] (Oldham 1886, 1893; Sharma and Bhadra 2014; Ministry of Water Resources 2016). Tectonic movements, water capture by other river streams and

---

[3] The existence of the Saraswati River remained largely confined to religious and anthropological documents for a long time. However, these beliefs are now validated. Geomorphological and remote sensing studies have confirmed traces of the river. The river is estimated to have originated in the Saraswati-Rupin glacier of the Himalayas. It used to flow through the plains of Punjab, Haryana and then in a south-west direction towards Rajasthan and Rann of Kutch in Gujarat. The Sutlej and

sand choking at origin are a few of the reasons behind the drying up of the Saraswati River. Drying of the main river also led to the reduction in the drainage areas of its local tributaries, the Ghaggar and the Luni. This led to the visibility of alluvial plains in different stretches. In addition to the alluvial plains, the change in course of drainage systems has also left water-filled patches across the desert (Dhir and Singhvi 2012; Kar 2014). Some of these patches are shallow and spread across a few kilometres. These patches store seasonal rainwater and then take the form of natural lakes. Mineral composition and high evaporation have made some of these lakes saline. The Sambhar Lake, the Lunkaran Sar, the Taal Chaapar and the Kavod are vast saline lakes in western Rajasthan. These lakes are widely used for salt and gypsum production. All these features help us imagine the past of the desert. These features, however, changed several times at different epochs largely due to tectonic movements and resultant climatic conditions. Tectonic movements have created hills and pediments (largely limestone, granite and rhyolite), pavements and Aeolian bed forms. In all, a lot of diversity exist in the desert. This diversity makes us to realize the variety of geomorphological and climatic processes that the landscape has undergone. One can also infer that each of these features has a story to tell, a story that involves interaction and networks between different natural elements. Examining these dynamics in detail lies beyond the purview of this paper. We will now examine how difficult living conditions are in the desert. This is with particular reference to low rainfall and water deficiency.

3   Rainfall Patterns and Droughts in Rajasthan

From the preceding section, we have a fair idea about the causes of deficit rainfall in Rajasthan. This idea becomes more reliable with the statistics presented in Table 1. The table gives an account of average annual and seasonal rainfall in western Rajasthan for the period of 1871–2016. In addition to western Rajasthan, the table also reports statistics for eastern Rajasthan and India as a whole. The inclusion of these two regions intends to highlight relative water deficiency features. For example, for the entire period of 1871–2016, western Rajasthan received an average annual rainfall of 249.07 millimetre (mm). This is almost 232 and 364% less than average rainfall received by eastern Rajasthan (578.08 mm) and All India (905.42 mm), respectively, for the same period. It should further be noted that this is but a cumulative figure for all administrative regions within western Rajasthan. A particular location may receive either much more or much less than this. Mrs. Dulari, 76 years old, resident of the Churu district, summarizes this spatial variation:

Baadal bhi thage, kabhi 10 kos pe to kabhi 100 kos[4] pe hage

Views of a local respondent

Yamuna rivers are estimated to be its tributaries. A large area of western Rajasthan was therefore drained by the river.

[4]Kos is a local unit for measuring distance. 1 Kos is approximately 1.12 miles or 1.8 km.

Table 1 Seasonal and average annual rainfall in western and eastern Rajasthan during 1871–2016

Period of the year	1871–1900			1901–1930			1931–1960		
	Rajasthan		All India	Rajasthan		All India	Rajasthan		All India
	West	East		West	East		West	East	
Jan–Feb	103.57	120.50	195.97	89.20	138.77	258.77	112.63	144.90	258.77
Mar–May	172.07	228.17	934.77	190.27	229.37	978.27	121.67	151.50	978.27
Jun–Sept.	2561.40	6581.07	8666	2441.43	5876.43	8789.40	2627.13	6797.67	8789.40
Oct.–Dec.	92.00	206.33	1157	104.57	263.60	1271.73	83.80	241.93	1271.33
Jan–Dec	2929.17	7125.45	10,963	2824.57	6508.40	10,637.5	2945.3	7336.07	11,307.2
Monthly average	244.10	593.79	886.42	235.38	542.37	942.27	245.45	611.34	942.27

Period of the year	1961–1990			1991–2016			1871–2016		
	Rajasthan		All India	Rajasthan		All India	Rajasthan		All India
	West	East		West	East		West	East	
Jan–Feb	85.37	110.20	211.53	92.81	125	204.92	96.82	127.95	230.17
Mar–May	252.83	230.17	919.87	238.58	243.27	970.27	193.89	215.76	206.37
Jun–Sept.	2585.43	6170.50	8388.40	2795.46	6305.77	8294.96	2596.88	6347.40	912.60
Oct.–Dec.	103.53	271.30	1199.23	124.23	257.73	12,020.92	101.01	247.92	344.42
Jan–Dec	3027.13	6782.33	10,719.1	3253.50	6931.96	10,673.12	2988.89	6396.98	10,865.10
Monthly average	252.26	565.19	893.26	271.13	577.66	889.43	249.07	578.08	905.43

Source Data from the Indian Institute of Tropical Meteorology (IITM) (2016)

Clouds also do mischiefs; sometimes they discharge at 11 miles, sometime at 112 miles

Translated views of a local respondent

In addition to spatial variation, there is also seasonal variation in rainfall. Western Rajasthan, like most Indian regions, receives maximum rainfall during July–September (Table 1)—that is, the monsoon season. This period accounts for almost 87% of the region's total annual rainfall in the era 1871–2016.

It has been observed that the Indian monsoon shows a trend cycle of 30 years of alternate dry and wet periods (Pant and Parthasarthy 1981; Guhathakurta and Rajeevan 2008; Shaban 2017). It is with this idea that Tables 1 and 2 have reported average annual and seasonal rainfall across thirty years periods: 1871–1900, 1901–1930, 1931–1960, 1961–1990 and 1991–2016. The idea is to observe any significant trend in rainfall in western Rajasthan relative to eastern Rajasthan and India as a whole. The 1871–2016, 1871–1900, 1931–1960 and 1991–2016 periods have been estimated to be wet periods. One can expect higher rainfall at these times. Figure 1 shows that the variation between wet and dry epochs is small in case of India as a whole. However, it is much higher in the case of both western and eastern Rajasthan.

The preceding sections give us a fair understanding that rainfall in Rajasthan is quantitatively low. Sometimes this deficiency reaches to such an extent that it becomes a crisis. Figure 2 plots the variation in average annual rainfall for each year with respect to mean annual rainfall for 1871–2016. The rainfall data shown in the graph reveals a normal distribution (although not a perfect one). Nearly 74% of annual rainfall lies between mean (M) and one-unit standard deviation (SD) above and below the mean, that is between M+SD and M−SD.

One can now make sense that recurring long spells of low rainfall make water a valuable resource in the region. Mr. Shamsheer Khan, resident of Churu, tries to interlink low rainfall and everyday hardships in his village through a local saying:

Chupadiya kaan, ho gaya snaan

Local saying

Wet your ears and your bath is over

Translated version

The above local saying may seem an exaggeration. But given the variations over time and geographic areas, the comparison makes us realize the extent of the crises involved. These crises had their own implications in the form for water shortages for drinking purposes as well as domestic use, agriculture and livestock. Both life and livelihoods used to suffer miserably during long drought periods (Fig. 3; Table 3).

Table 2 Standard deviation (SD) and coefficient of variation (CV) for seasonal and average annual rainfall during 1871–2016

Period of the year	1871–1900			1901–1930		1931–1960		
	Rajasthan		All India	Rajasthan	All India	Rajasthan	All India	
	West	East		West	East		West	East
Jan–Feb SD	91.90	104.95	98.45	99.45	123.40	137.52	123.36	132.91
Jan–Feb CV	0.89	0.87	0.50	1.11	0.89	0.50	1.10	0.92
Mar–May SD	116.00	180.49	210.86	186.89	219.99	205.11	96.53	123.71
Mar–May CV	0.67	0.79	0.23	0.98	0.96	0.22	0.79	0.82
Jun–Sept. SD	865.52	1599.09	908.09	1238.71	1940.48	795.85	835.15	1606.44
Jun–Sept. CV	0.34	0.24	0.10	0.51	0.33	0.10	0.32	0.24
Oct.–Dec. SD	114.58	23,057	351.07	147.34	287.99	406.81	104.79	363.20
Oct.–Dec. CV	1.25	1.12	0.30	1.41	1.09	0.35	1.25	1.50
Annual SD	75.29	129.15	90.84	117.13	176.00	80.42	74.34	144.55
Annual CV	0.31	0.22	0.10	0.50	0.32	0.09	0.30	0.24

(continued)

Table 2 (continued)

Period of the year	1961–1990				1991–2016				1871–2016			
	Rajasthan		All India		Rajasthan		All India		Rajasthan		All India	
		West	East	West		West	East	West		West	East	West
Jan–Feb SD	104.33	87.30	112.45		112.42	108.91	133.39		919.97	101.89	120.17	113.28
Jan–Feb CV	0.40	1.02	1.02		0.53	1.17	1.07		0.45	1.05	0.94	0.49
Mar–May SD	208.94	267.04	205.11		216.85	170.41	184.50		197.78	181.84	185.48	206.37
Mar–May CV	0.21	1.06	0.89		0.23	0.71	0.76		0.20	0.94	0.86	0.22
Jun–Sept. SD	1059.57	1063.86	1479.85		944.34	895.65	1442.10		762.46	986.62	1632.35	912.60
Jun–Sept. CV	0.12	0.41	0.24		0.11	0.32	0.23		0.09	0.38	0.26	0.11

(continued)

Table 2 (continued)

Period of the year	1961–1990			1991–2016			1871–2016			
	Rajasthan		All India	Rajasthan		All India	Rajasthan			All India
	West	East	East	West	East	East		West	East	East
Oct.–Dec. SD	359.99	123.23	307.22	271.20	207.62	336.01	341.25	140.76	303.08	344.42
Oct.–Dec. CV	0.28	1.19	1.13	0.23	1.67	1.30	0.28	1.39	1.22	0.29
Annual SD	105.99	96.58	122.66	92.06	82.24	123.00	73.64	90.28	140.60	90.61
Annual CV	0.11	0.38	0.22	0.10	0.30	0.21	0.08	0.36	0.24	0.10

Source Data from the Indian Institute of Tropical Meteorology (IITM) (2016)

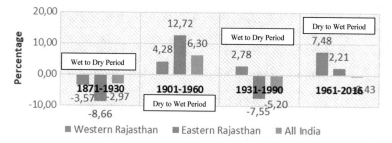

Fig. 1 Percentage variation in annual rainfall in wet and dry epochs (1871–2016). *Source* Data from Table 1

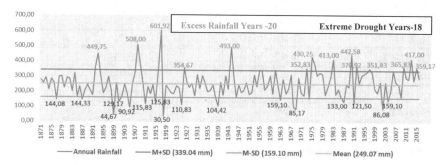

Fig. 2 Variation of average annual rainfall in western Rajasthan during 1871–2016. *Source* Data from the Indian Institute of Tropical Meteorology (IITM) (2016)

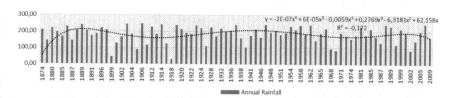

Fig. 3 Recurring rainfall deficit and drought years in western Rajasthan during 1871–2016. *Source* Data from the Indian Institute of Tropical Meteorology (IITM) (2016)

Table 3 Prolonged rainfall deficit and drought spells in western Rajasthan in 1871–2016

Consecutive years with average annual rainfall less than mean	Extreme drought years in between	Average annual rainfall (in mm)
1885–1891	1887	196.38
1901–1906	1901, 1905	177.28
1911–1915	1911, 1915	179.58
1918–1925	1918, 1925	178.85
1935–1939	1935, 1939	182.47
1946–1949	Nil	202.94
1957–1958	Nil	225.58
1962–1963	1963	193.33
1968–1969	1968, 1969	90.88
1971–1972	Nil	183.63
1980–1981	Nil	205.92
1984–1989	1987	195.69
1999–2000	Nil	205.54
2004–2005	2004	171.71

Source Data from Indian Institute of Tropical Meteorology (IITM) (2016)

4 Understanding Nature Through Collective Efforts and Water Conservation[5]

Why did people still choose to live here despite difficult environmental conditions, especially with regard to water? Historically speaking, given the desert-like conditions, locals had a choice to migrate to more resourceful regions. Yet they did not. There are few communities such as Gadoliya, Kalbeliya, Roma, Bhopa, Bhils and even some Gujjar and Gadar clans who decided to migrate. But even they moved largely within the region alone. Also, their movements were seasonal. Reasons for such regional adherence were simple. First was the sense of belonging to the native land. This feeling was also accompanied by a sense of insecurity and fear to settle and prosper in a new place. Geographical escape, according to the locals, was not a permanent solution. But this does not mean that people felt helpless and decided to simply suffer nature's hardships.

Instead, they reverted to nature as their saviour. This idea may seem awkward. But locals here believe that nature always has something good to offer everyone. They draw an analogy between mother and nature. A mother can be upset and angry.

---

[5]The views in this section are largely an outcome of a field trip conducted in western Rajasthan in April–June 2017. Fieldwork involved visits to towns and villages, informal discussions with villagers, academicians, and government officials. In all, three focus group discussions and 25 detailed interviews were conducted.

However, she can never harm her children. On the contrary, she will make them happy by giving whatever she has. Nature also works on a similar note. It always offers resources so as to fulfil the basic necessities of life. This is true even in regions of scarce water. Resources and ways to deal with water scarcity are present in the surroundings. However, these resources are not available in ready-to-use form. One has to make efforts to explore these resources and create conditions to use them. In short, one has to first understand the way nature works. This working of nature is best demonstrated by local flora and fauna, which have somehow learned the art of adapting to the water conditions. Locals realized that they had been lagging in this aspect.

Inadequate understanding of nature can have many causes. One of the reasons locals identified was the absence of collective effort. Collective effort was much needed so as to explore natural features and processes so as to access water more efficiently. This however was not easy. Indeed, it was difficult as nothing exists in isolation. Natural features and processes are all interconnected. For example: vegetation, soil, wind movements, temperature and landforms are all in continuous interaction. Also, each of these categories has a lot of diversity and dynamics within it. An overall understanding cannot be avoided. These features together determine rainfall patterns and thereby the water situation. All these considerations made the exploration task of nature quite complex. No doubt, it was beyond the capacity of any one individual or community. Locals realized these constraints and therefore decided to work together so as to share knowledge and collectively devise methods to survive. All these aspects are best understood by few lines that can be inferred from Mishra's phenomenal work The Ponds are Still Alive (1993):

> ...Maru mein jeena kathin hai, lekin ye jagah abhishaap nahi Varsha kam hai, lekin aisa nahi ki bilkul nahi Rehne ke liye prakriti ko samajhna tha, ek paheli hai jise bhoojna tha Sab log ek saath aaye, kaise rehna hai ek dusre ko bataye Jeevan sanjona aane laga, paani sab kuch hai aur sab kuch paani se ye samajh mein aane laga...

> Maru (Part of the Thar desert also called Land of the Dead) is difficult to live in, but this place is not cursed. Rainfall is scanty, but this does not mean there is no rain. In order to live, nature has to be understood, a riddle has to be solved. Everyone came together, life itself was shared by all. Life began; that water is everything and everything is because of water was understood by all.

Translated version

Locals keenly observed nature. But observing nature was not a task to be done in isolation. That is, one does not have to allocate time each day exclusively for it. Instead, it was seen as an integral part of life in general. Individuals continued their everyday activities as they used to—the only difference being that now more attention was paid to nature. This was both during work and leisure times. How does nature affect an activity? How does an activity affect nature? How are natural processes interconnected? And most importantly, how can more water be ensured? All these were usual considerations. Answers to these considerations however varied from one individual to another. This variation was not just due to subjectivity in observations, but also due to differences in the roles and activities in which people were engaged. Nature was therefore explored through various aspects. All these individual observations were shared with others and with the community as a whole. They were shared not so as to showcase one's own intelligence or skills of minute observations, but so as to have input from others—inputs which may prove one's inference to be wrong—or may enhance it by additional information. Such knowledge sharing also brought communities together. This is important given the practices of cultural and class differentiation in the region (The Rajputana Gazetteer 1879; Rudolph 1963). Indeed, for this task everyone agreed to contribute and learn from each other (Mishra 1993).

The collaborative effort to learn about nature did not involve only adult citizens—it was an intergenerational endeavour. That is, children were not seen as inexperienced and unfamiliar—they too contribute with questions on different aspects. And sometimes children would infer something which adults could not think of. This shows that anyone could be an educator, irrespective of age. Such an active participation by children was not by chance—it was largely because of an upbringing centred around nature. Mothers and female community members must be largely credited for this. They are the child's first teachers. Given their everyday engagement with diverse activities, female members develop insights on different natural elements. All these insights are shared with children as they grow, play, work, eat and discuss anything. This sharing of observations, again to re-assert, was not done purposely. It was the way of life itself placed knowledge of nature as basic common sense (Mishra 1993). This common understanding of nature strengthened with time. More observations were made, earlier observations were refined, and new aspects were touched upon. This helped locals develop more knowledge of their surroundings.

Such an extensive and prolonged effort to understand nature gradually helped locals to figure out how to deal with water deficiency. Locals realized that they cannot control the rain. So, in order to deal with water deficiency, they cannot rely on rainfall alone. Also, there were no extensive water bodies in the region depend upon. There were, however, a few naturally occurring small water pits and patches, from which locals used to fetch water for both drinking and domestic purposes. These water pits became the solution to water deficiency.

> ...Varsha yaha pe kam hi hai aur bade jal strot bhi nahi the. Lekin thode chote prakritik jal strot the, jaise ki gadde, kahi chote kahi bade. Inhi gaddo se roz peene aur dusre kaamo ke liye paani liya jaata tha. Ye gadde bhi har jagah nahi the...ghar ki aurato ko 10–12-kilometre tak chalna padta tha inse paani bharne ke liye. Paristithi kathin thi. Lekin jal samasya se nipatne ka upaay bhi inhi gaddo se aaya. Inhi ko baar dekhkar, garmiyo mein suka dekhkar aur baarish mein phir bhar jaata dekhkar aur inse paani nikaalkar laga kya hum prakritik jal strot jaise bhandaare nahi bana sakte kya? Kya hum khud gadde nahi khod sakte...kya inn khode hue gaddo mein paani roka ja sakta hai...bas yahi wah shan tha jab Maru ki kismat ko purvajo ne badalne ka soch liya...
>
> Mr. Ghanshyam, 76 years old, Local Respondent

> ...Rainfall has always been scanty; there were also no large water bodies. However, there were relatively small natural water bodies – pits - some small and some large. Water was fetched from these pits for drinking and other purposes. These pits were not located everywhere.... women used to walk nearly 10–12 kilometres every day so as to fetch water from these pits. It was a difficult situation. But the solution to deal with water deficiency also evolved with these pits. Watching these pits evolve - watching them dry in summer, fill again

with rainfall - and withdrawing water from them made us ask ourselves whether we too, like nature, could construct such water reservoirs. Can we ourselves dig pits... can water be stored in these constructed pits... this was the moment when our ancestors committed themselves to change the destiny of Maru (Thar)...

Translated version

Water pits laid the foundation of water conservation in the region. The idea was drawn from nature and validated local beliefs that solutions to natural problems are available in the surroundings. And sometimes large-scale, complex problems may have simple solutions. It all depends on the way we engage with the problem and with our surroundings. Of course, just having the idea of a water pit did not by itself relieve the region from water deficiency. Where should pits be dug? What should be their size? And what can be done so that these pits hold rainfall water for long periods of time? These were the considerations locals have struggled with for a long time. Pits dug randomly would not guarantee water fill and storage. Patience, determination and continuous experimentation helped locals to develop overall strategies taking into account all the elements involved. These included considerations regarding slopes, soils, depth, and rainfall probability at chosen construction areas. Taking these into account, locals gradually evolved pits of varying size and designs. The idea behind these pits was not to waste a single rainfall drop—conserve all possible water. Instead of allowing rainfall water to infiltrate the ground, it was to be adequately stored and used cautiously.

The succeeding sections discuss a few such water structures. Reminding the reader, again, that these water structures are not just about water. They represent a culture—a culture that placed nature and water at its centre.

5   Water Conservation Structures

5.1   Taankas

The *Taanka* or tank is an underground reservoir used to store rainwater drainage from either natural or man-made catchments (Vangani et al. 1988; Mishra 1994; GRAVIS 2003). The stored water mainly serves drinking purposes for a single or small group of families. *Taankas* provide between 4 and 8 months to a year of drinking water. They also serve as storage units for transported water once the conserved rainwater is exhausted. Storage of water in *taankas* ensures taste and cleanliness; it also saves time, energy and money for there is no need to fetch water from faraway places. The *Taankas* are traditionally found within the larger premises of the house or in the courtyard of many houses in western Rajasthan. They are often beautifully decorated with tiles, and *tulsi* plants are grown around it to keep the water cool, clean and healthy. The characteristics of a *taanka* are as follows:

(i)   *Shape*: Taankas can be circular, rectangular or square in shape, circular being the most common. They are constructed at a height so as to maintain the cleanliness of drinking water and prevent polluting water seeping in. They are usually 3–4 m in diameter and 4–5 m deep. In circular taankas, the depth and diameter are the same. The size and capacity of a taanka is determined by the number of families dependent on it, variations in lifestyle, requirements for livestock, nurseries and any additional consumption. The capacity of a taanka is usually around 25,000–50,000 L. Taankas are primarily made of two materials: local materials such as lime and cement plaster give it a lifespan of around 3 years; cement and concrete ensure a lifespan of 10–15 years. The lining of these materials helps keep evaporation and leaking under check. Masonry taankas also exist, although uncommon. Regular cleaning, desiltation, repair and maintenance are essential for a functioning taanka. The mouth of the taanka is covered so as to prevent dust and reduce water loss by evaporation. A small lid-covered opening allows fetching water from the taanka with the help of a bucket tied to rope.

(ii)   *Catchment*: The catchment is an essential component of the taanka, wherein rainwater is collected and runoff is generated. Sizes of both naturally occurring and artificially constructed catchments vary between 15 and 25 m in diameter. They are provided with a gentle 1–2% slope and surface qualities that determine the runoff coefficient (ratio of runoff to rainfall in a catchment). The water from catchments flows along the slope and is collected within the taanka. In present times, rooftops of buildings serve as the best catchment devices and usually provide a personal ownership of taankas or ownership by a few households. A rainfall of 50–75 mm is sufficient to fill the taankas. Physical properties of the catchment determine the runoff rate/quantity of runoff; chemical characteristics such as the presence of fertilizers or pesticides determine its quality. The surface of the catchments is made smooth, compact and semi-impervious with a layer of the silt (obtained while cleaning ponds) or wax, asphalt, bitumen and bentonite and compacted by a local method of rolling with *Crotalaria burhia* (a herb in the Thar with soil binding properties) and sand. Soil and dune sand are waterproofed with murrum (a gravelly laterite material), cement, lime, gypsum and stone crust. Large boulders, stones and unproductive vegetation are cleared so as to ensure unhindered runoff. These arrangements are meticulously repeated every year at the onset of monsoons so as to ensure an efficiently functioning taanka.

(iii) *Inlet and silt catcher*: The inlet from the catchment into the taanka is protected by iron sieve/mesh so as to guard any particles, silt, soil and clay from entering the taanka.

(iv)   *Outlet*: Approximately a 30 cm × 30 cm outlet is provided at the opposite end of the inlet so as to release excess water from the taanka into a common outlet such as river, pond or other water bodies. This is also guarded by an iron mesh and keeps away insects, rodents and garbage.

Taankas are a good solution for the drinking water problems in western Rajasthan. They are highly economical to both build and maintain. They reduce the stress on women to fetch water from long distances and help save money to fetch marketed water in times of drought. They also provide employment to rural artisans with masonry skills; they are cheap to build and do not require specialized tools for construction. Taankas not only improve standards of life among desert dwellers—especially in areas with scarce groundwater—but also prevent depletion of groundwater itself.

The history of water conservation practices in Rajasthan reveals that taankas, although an age-old practice, came into prevalence much later than ponds and baoris (stepwells). The first-known construction of a taanka in western Rajasthan was that by Raja Sursinghji in the village Vadi ka Melan in Jodhpur during the year 1607. Later, a taanka in the Mehrangarh Fort, Jodhpur, was constructed during the regime of Maharaja Udaisinghji in the year 1759 (Vangani et al. 1988). During the great famine in 1895–1896 (also known as mahakaal), taanka construction was taken up on a wider scale in this region. The largest taanka was constructed at the Jaigarh Fort in Jaipur—it holds 30 million litres of water. The construction of taankas was considered a religious service by both the kings and the people. Taanka water was used for worship in the temples. Almost all forts and palaces had taankas, some in secret locations and intended to serve the army. Most houses had started building their own taankas, the number and size of which often were a symbol of status. Western Rajasthan has taankas of sizes varying from 1000 to 500,000 L, according to local needs. The Taragarh Fort in the Bundi district has taankas much praiseworthy for their design and capacity. Several taankas have been built across the Thar with new, improved designs by both the government and different NGOs.

Improvement in design and Mass Construction: Central Arid Zone Research Institute (CAZRI), Jodhpur, invested 40 years of work and research so as to improve the design and construction of taankas and make them cost-effective. The improved taankas are commonly a stone masonry with plaster and cement concrete. They provide for three inlets and an outlet. Water is fetched manually with a rope and bucket or through a hand pump. These improved taankas by the CAZRI have a lifespan of over 20 years and have become both very popular and widely adopted in western Rajasthan. About 12,000 such taankas have been constructed in western Rajasthan under the Rajiv Gandhi Drinking Water Mission intended to store 475,000 m^3 of water and serve 32,000 people throughout the entire year (Goyal and Issac 2009). GRAVIS, another NGO, has been working to construct taankas for the neediest households. Village development committees select dhanis and households based on factors such as lack of water storage facilities, livelihoods below the poverty line, travel of 1.5 kms or beyond so as to fetch water, and households consisting of widows and elderly people with no family support. As of 2017, GRAVIS has energized the locals so as to build about 7000 taankas. In 2014–2015 alone, the organization helped build 281 taankas (GRAVIS 2018). Several NGOs organize awareness programmes by bringing together villagers, government organizations and other NGOs so that everyone can contribute to the improvement of these water conservation methods. Villagers themselves have in many regions taken the initiative to build taankas. One standing example is Ismailpur, a small village in Jhunjhunu with 145 households. All the houses in this village have a 13-feet deep underground tank with a storage capacity of 20,000 L. Every member of the house uses a quota of 8 L per day—with a discipline that allows water to be sustained for a year. This has inspired around 20 other neighbouring villages to construct household taankas. The number of households having personal taankas is now increasing in western Rajasthan.

5.2 Khadins

The khadin, also known as dhora, johad or check dam, is a traditional irrigation-free, rainwater-based farming system. The khadin is basically a low-lying area wherein rainwater flows down naturally (IRC 1988; Mishra 1993). It is an integration of native dryland farming and a highly efficient moisture conservation system. The khadin is a runoff agricultural system that works on the principle of harvesting rainwater from a natural upland catchment and directing it to low-lying farmlands so as to conserve moisture in the soil and facilitate crop production. It is a compact, watertight system of rainwater harvesting wherein water from uplands saturates the agricultural plains into a muddy basin of alluvial silt and clay. Upland water keeps the soil moist and prevents water evaporation during summers. This method is extensively practised in regions of high water salinity in the extreme west of the Thar Desert, particularly in and around Jaisalmer and Jodhpur, and ensures at least one successful crop per year for a sustainable economy. The major components of the khadin system are as follows:

(i) water collecting areas or catchments—landscapes of rocky and gravelly outcrops lying at naturally elevated levels where rainwater is collected, (ii) contoured bunds (channels) which run along the rocky gravel so as to direct the runoff to the khadin farms (agricultural farms) with minimum seepage loss, and (iii) moisture storage basins (khadin farms) in low-lying plains of porous soil for crop cultivation. The basement of khadins is a rocky surface upon which gradual deposition of sand, silt and clay increase the fertility of the land. This maintains a supply of natural moisture and nutrients to the crop roots. Usually, the khadin area is 10–14 ha and a ratio of farmland to catchment area is maintained between 1:6 and 1:18 so as to ensure uniform moisture supply. A rainfall of 75–100 mm is enough to charge the soil with sufficient moisture. (iv) Earthen and stone khadin bunds or embankments are a main feature of the khadin. These check dams are usually 100–300 m long and constructed along the plain so as to arrest and conserve the runoff, leading to improved percolation and recharge. This is strong enough to withstand the force of inundated water and flash floods. (v) Sluices and spillways are provided beneath the bund so as to remove excess water from the farmlands when required, and (vi) villages (earlier, Paliwal villages) are located at the peripheral zones of the khadins. The moisture content in the farmland recharges groundwater and facilitates vegetation further downslope below the sluice gate, supporting community settlements.

Utility: Crop cultivation in the khadins is an unusual method, wherein crops grow in reserved farm water without any flow channel irrigation. This means that khadins require time for water recharge as well as a specified time for crop cultivation. During the rains (July–September), the khadin farms are seasoned for cultivation. The rains saturate the farms, layers of mud and clay accumulate, and livestock is grouped near the mud beds so as to enrich land fertility with their excreta. In late autumn and early winter (October–November), the lands are ploughed and seeds are sown. The crops, mainly millets (wheat and chickpea), grow throughout the winter and are harvested in March. Presently, woody perennials and fruit-bearing trees such as the neem, babul, khejri, ber, date palm, and lasura are also cultivated in khadin farms and along the bund so as to prevent soil erosion. They provide green fodder to cattle; the khejri trees also help improve soil quality through nitrogen fixation (Prasad, et al. 2004). Khadins, also referred to as man-made oasis, are the only land areas where winter crops are grown under rain-fed conditions.

Improvement in Design: Traditional khadins were good cultivation systems. However, certain drawbacks in their architecture led to insufficient utilization. They had no water outlets, for example. Uneven land surface and runoff led to improper water distribution in the farmlands, waste of excess water, disproportionate catchments and even accumulation of salts in and around the khadins. The present-day, improved khadin structure and design keep these problems under check and increase crop productivity.

Apart from improving agricultural yields and providing financial security to locals, khadins make positive environmental contributions. The compact fields prevent soil erosion as well as the washing away of minerals and manure—thus plant and crop remains are naturally decomposed in the fields and much increase fertility. About 62% of khadin water contributes to groundwater recharge (Khan and Narain 2003) and facilitates vegetation growth around the site. This also provides fodder for animals and shady trees for perch and rest. A study conducted by the CAZRI, Jodhpur, showed that about 2.5–3.0 t/ha of wheat and 1.5–2.5 t/ha of chickpeas could be successfully cultivated in khadins without the use of chemical fertilizers (Kolarkar 1990). The khadin at Baorli-Bambhore in Jodhpur generated an income of Rs. 18,500 ha^{-1} through food production during the extreme drought of 2002 (CAZRI 2007). Improved agronomic practices in the Baorli-Bambhore khadin improved the grain and straw yield of pearl millet crops by 20.19% and 12.15%, respectively, in the year 2013 and 20.68 and 20% in 2015 (Sharavani 2016). Khadin lands provide twofold to threefold increases in crop production as compared to normal agricultural fields (CSR Portal 2018). The khadin agricultural system is a very old practice; it witnessed a downfall but now rose again and is now much benefitting crop production in the arid Thar.

Khadins have been in existence for over 500 years. Similar structures can also be found in the Middle East and are estimated to be centuries-old (Planning Commission 1981). In the case of the Thar, khadins were first developed by the Paliwal Brahmins of Jaisalmer in the fifteenth century (The Rajputana Gazetteer 1879; Mishra 1994). Paliwal Brahmins were prosperous cultivators who developed an integrated system of catchments and farms for crop production, even during the worst droughts. Khadins were the main source for income for the Paliwal community. Probably instigated by their success, Paliwal cultivators faced constant challenges from occupational rivals for diverting runoffs and were irrationally accused for recurrent droughts. Finally, in the nineteenth century the uprise by Rajput rulers in Jaisalmer against the Paliwals led the community to abandon their villages and khadin lands. Discontinued use and neglect led the khadins to a state despair. It is only in the second half of the twentieth century that the value of khadin cultivation had been recognized again and both restoration of old and building of new khadins has been initiated (IRC 1988). The Rajasthan state government in cooperation with several NGOs and even individuals has been working towards large-scale repair and development of the khadin system of cultivation. New designs of farms are being adopted so as to optimize water infiltration, runoff routing and soil water storage capacity.

Both central and state governments have taken various initiatives so as to both build new and improve the conditions of existing khadins in Rajasthan. The initiative started since 1965, after the government restored 66 khadins in Jaisalmer. The government of Rajasthan and the state irrigation department have taken over the responsibility of repairing and maintaining these structures. The drought-prone area program (DPAP) was the earliest area development programme established by the central government in 1973–1974. The state government provided financial subsidies for building and renovating khadins. Numerous NGOs also started working in different regions across Rajasthan so as to make khadins available for use. In 1985–1986, during a severe drought in the Alwar district, Tarun Bharat Sangh—an NGO led by Rajendra Singh—revived the culture of khadins and revolutionized water conditions in Alwar (Tarun Bharat Sangh 2009; Sinha et al. 2013). Depending on the administrative authority, the khadins are leased out to cultivators—often large communities or a smaller number of families. New khadins are being built under the desert development programme (DDP). The Gramin Vikas Vigyan Samiti (GRAVIS) has built 120 khadins to the benefit of 360 families in the year 2015 alone (GRAVISINDIA 2017a, b). During its entire time span of work, GRAVIS has led villagers to construct over 5000 khadins across Rajasthan (GRAVISINDIA 2017a, b). Natural resource management (NRM) for watershed management constructed 157 khadins in Barmer and Baitu during 2014 and 2015, and many more constructions are in progress which will be completed in August 2018 (CSR Portal 2018). The CAZRI developed a khadin of 20 ha area in Baorali-Bambore in the year 2002 that helped farmers grow both kharif and rabbi crops. During 2002, the CAZRI applied technological improvements and enhanced land productivity in a small khadin named Darbari Khadin with a catchment area of approximately 4 km^2, located 15 kms away from Jaisalmer (Narain and Kar 2005). An NGO named the Aakar Charitable Trust (ACT) founded by Amla Ruia played a pivotal role in changing the water and economic conditions of several villages in Rajasthan. The ACT constructed 200 check dams in 100 villages and generated an income of 300 crores for two lakh villagers during the years 2000–2015. This has much improved economic conditions among villagers, who now reap three crops a year and earn additional means from animal husbandry (The Better India 2015; Peerzeda 2017). The villagers themselves have garnered motivation to construct more khadins.

5.3   Naadi/Ponds

'Naadi' is the local name for village ponds—which are usually manually dugout water bodies. They are the oldest as well as most prevalent and widespread water conservation strategy in the Thar Desert. Naadis are a community-based drinking water system for livestock, humans and wildlife. Sometimes they serve as supplementary water source for growing vegetables, nurseries and horticultural crops. A small naadi is also known as naada; a large naadi is known as a talab. Most villages in western Rajasthan have around three to five such structures of various sizes. Rajasthan is filled with ponds and has been so throughout history.

The first recorded naadi in Rajasthan is the Jodhnadi in Jodhpur, built in 1458 AD. Jodhpur also had the first stonewalled naadi. It was constructed in 1520 AD during the regime of Rao Jodhaji (Saxena 2017). Constructing a naadi was a symbol of prosperity and also a sacred deed. Rajput rulers constructed a number of naadis during their rule. The Bal Samand Lake in Jodhpur, for example, was built by Balak Rao Parihar in 1159 AD so as to address the water problem of Mandore. The Gadsisar Lake in Jaisalmer was constructed in 1367 AD by Raja Rawal Jaisal, the first ruler of Jaisalmer. It was the area's main water source. The Kalyana Lake in Jodhpur was built by Pratap Singh in 1872 AD and still is the city's main water source. The Rajsamand Lake in Rajsamand, the Anna Sagar and Pushkar lakes in Ajmer, the Kailana in Jodhpur, the Doodh Talai, Fateh Sagar, Pichola and Jaisamand are but a few of the lakes in Udaipur—the city of lakes. Over a period of time, while some of the old naadis such as the Range-ki-nadi and Pritaro-nado (about 450 years old) have dried up completely due to negligence, many naadis are still functioning and have grown to be the most common water conservation structure, dotting entire western Rajasthan. A survey by the Central Arid Zone Research Institute (CAZRI) found that the Nagaur, Barmer and Jaisalmer naadis date from 1436, 592 and 1822, respectively (Saxena 2017; Vangani et al. 1988). The Jaseri Lake near Jaisalmer has a history of never getting dry. The recent census of these structures is however not available.

The structure of a naadi comprises a large-sized catchment area (also known as a paal or agor) in the downslope wherein is situated the pond/naadi. The pond is excavated into a pit so as to hold runoff from the catchment. During the rains, the catchment fills and the overflow drains down into the pond and stores sufficient water. Often when the amount of rain is not enough to fill up the naadi, the water released from the catchment helps keep the naadi full and the water accessible for use. The storage capacity of a naadi is dependent on its area of location and on the physical characteristics of the adjoining catchment and runoff. The size of a naadi varies from small (with a capacity of around 700 m^3) to large (around 20,000–40,000 m^3 of water). The size of a naadi in the dunes is usually small with a small catchment area and less runoff as well as greater loss through seepage. The ones in the sandy plains are usually larger, with a bigger catchment area and less loss through seepage. After the rainy season, the water in naadis can last for two to six months until up to a year in the big ponds. During pit construction, the dug-up soil is placed around the periphery of the naadi on three sides (to form what is known as a bund) so as to prevent soil erosion and waste of water through runoff. The fourth side is connected to the catchment area via a plain slope of land. The water from the catchment is directed into the naadi usually via small streams. This runoff contains a lot of particles like stones, gravel, soil and sand. A barrier (also known as khura) is placed upstream of the pond by creating a stone network (called chhedi) that sieves out these contaminants so that only the filtered, clean water flows into the pond. This stored water serves the community and is also indirectly useful to improve water conditions in the state.

Water stored in the naadi is beneficial for groundwater recharge through seepage and deep percolation. They recharge nearby wells and beris. This makes water available in these structures for a longer period of time. Sufficient groundwater recharge also facilitates availability of water in public tube wells. Stored water in the naadis benefits nearby farms by aiding growth of crops, grass and various trees, products of which can be used for human consumption, animal fodder, as medicines or as source of income. These have also facilitated the abundance of flora and fauna. The ponds are community structures, and the entire community played a major role in its construction, maintenance and use.

Over time, wind-blown sand and contaminants from the runoff accumulate at the bottom of the pond (a process called siltation) and reduce its capacity. Periodic desilting (removing the accumulated soil or silt) is hence carried out. The channels from the catchment to the pond are also cleaned so as to maintain the runoff rate. Earlier, Amavas (moonless night) and Poornima (full moon night) were dedicated times for public service (traditionally a time of withdrawing from personal engagements), in which communities would engage in maintaining the naadis. The naadis are not only a source of water but a part of culture and social norms. They are often worshipped and considered sacred. The festival of Sawan Teej or Shravan Teej is celebrated so as to welcome the first rains during July/August, when women flock to the naadis in groups and offer prayers to the Goddess Parvati and Lord Shiva. These festivals usually mark a time for the community to get together, clean the ponds and offer prayers for rains. This builds intimacy and compassion among the villagers. All members of the community and those living away from it, irrespective of caste, maintain and use the naadi. The community also laid down certain rules and regulations so as to maintain water quality in the ponds. Livestock were not allowed to graze and disturb the soil in the area around the pond and catchment. Shoes were also not allowed. Both humans and livestock were not allowed either to enter, urinate or defecate in the area. There were also strict restrictions against fetching water directly from the catchment. However, these rules are no longer practised in many parts. Poorly maintained naadis suffer water losses due to seepage, evaporation and pollution due to both human and livestock defecation which has propagated guinea worms, water hyacinth, mosses and algae in many places.

Digging ponds is laborious activity and an economically prudent for locals. Earlier, building ponds was a way of life in among desert communities. However, now it is a necessity compelled by water crises. The lack of specialized knowledge needed to construct it is helping communities unite. Sometimes, the state (or ruling Raja) played a role in building, maintaining and regulating some of the big ponds and collecting tax and revenue for maintaining them. In current times, in order to meet the needs of a growing society, the government, several NGOs and even individuals have come forward to help out with equipment, strategies and new concepts for building, cleaning ponds and maintaining water quality. These ponds also provide some locals with a source of income. The CAZRI has designed numerous naadis with low-density polyethylene lining (LDPE) intended to reduce water loss through seepage and evaporation and can sustain water for a longer time. In recent years, this technique has been used by Rajendra Singh in the Tarun Bharat Sangh so as to restore water conditions in numerous villages across western Rajasthan. Shashank Singh Kachwaha, a mechanical engineer, helped rejuvenate water conditions in the Chota Naraina village in Ajmer. Although initially at a slow pace, the number of ponds in western Rajasthan is ever increasing—propelled by local determination and by increased support from individuals, NGOs and often the government.

5.4   Beris

Beris or kuis are earthen, pitcher-shaped, shallow wells (Mishra 1994; GRAVIS 2003). They store rainwater percolation and provide sweet drinking water. Beris usually exist near catchments or taankas in dry areas within the extreme west of Rajasthan—which suffers from high groundwater salinity, especially in Bikaner, Jaisalmer and Jodhpur. When khadins, naadis and taankas dry out, their water percolates into beris. In the sandy terrains, beris assimilate water from deep underground, which ensues even after a mild rain. The layer of soil overlying the deeper, brackish aquifers contains pockets of sweet rainwater. Old alluvium soil comprising sand, gravel and rock creates these pockets of filtered rainwater, which is sweet in taste. Traditionally, these stores of sweet water are called paar and can be identified by vegetation growth on their surface. Beris are dug out at these places. The trapped, slowly percolating, sweet water of the soil is colloquially called rejwani or rajani pani. Water conservation through beris by constructing a number of them together at one place is known as the paar system. Usually, there are six to ten beris in a paar, but depending on the size of the paar, the number of beris can vary from less than six to twenty. The rainwater harvested by the paar system is also known as patali pani.

The beri resembles a well in its structure, but it contains sweet water from much above the brackish water layers providing water from wells. The regions where beris are dug up are strategically chosen—near catchments, where water naturally percolates and finds its way into the beris. Usually, good vegetation growth—especially of neem, khejri and rohira trees—marks the best-suited location to get sweet water (Mishra 1994; GRAVIS 2003). The sand and soil are dug up till a layer of clay, shale or gypsum is reached. Traditionally, this hollow underground cistern is not additionally lined; instead, a layer of plaster is applied, and the impervious stratum prevents water loss through seepage. The composition of this impervious layer varies from region to region. In Jaisalmer, it is a layer of bentonite, while in the neighbouring Barmer district it is a mixture of sand, gravel and rock. The pitcher shape of beris ensures a small mouth opening so as to reduce evaporation and provide a broader base for water to seep in through a large surface and store enough water for 4–5 families to use for several months. Even in times of minimum rainfall, the science behind beri construction allows it to be replenished and serve as a dependable source of water, even if in small quantities. Depending on the size, the beris serve as storage for 100,000–400,000 L^3 of water. Usually masonry structures, the beris can be either temporary or permanent—dug up according to requirements. The temporary beris dug up near khadins or naadis and intended to last for a season are about 2–3 m deep; permanent ones are usually 7–10 m deep. If the water is exhausted, the beris are dug deeper. Beris have a 1–2 m diameter. The mouth of permanent beris is lined with shale so as to provide stability and covered with a stone slab so as to prevent water evaporation, infiltration of dirt and dust, and perching over by birds. Functional beris have this stone platform with a hole so as to make it convenient to fetch water.

Traditional stonewalled beris are more durable—and some of these age-old beris are still functional. About five to six decades ago, the traditional beris stopped being constructed due to the lack of expertise. Yet recently its significance in sustaining desert life has been recognized all over again. Cement has become a regular material, and numerous cement beris are being constructed across western Rajasthan. Two concentric iron sheets are erected, and the space between them is filled with cement, after which digging is carried out. The beri is a relatively cheap method for harvesting water, costing about INR 3000–30,000, depending on the size (GRAVIS 2003; Jhunjhunuwala 2005). Villagers have learnt the craft of building beris and pool money so as to invest in constructing community beris themselves. Several households also have their personal beris. Family beris are constructed mainly by the members of the family so as to save money, and whatever is invested in the process is regained back as economic or lifestyle profit for the family. At times of sufficient rainfall, large beris not only provide drinking water, but can also support enough water to irrigate lands. Beris have also benefitted from positive intervention from the government and NGOs.

Government agencies and several NGOs have taken up a keen interest in reviving the status of the paar system in Rajasthan. The Thar Integrated Social Development Society (TISDS), a Jaipur-based NGO, was funded by the Centre for Science and Environment (CSE) so as to revive and improve the status of beris in western Rajasthan. Jethu Singh Bhatti, the general secretary of TISDS, initiated this mission of paar as a drought-proofing model in 2003 in the Manapia village in Jaisalmer. This construction programme not only provided sources of income for the locals but also led to great gains in terms of water status in the village (CSE 2006). The GRAVIS, along with the Regional Housing Development Centre and the Jodhpur University, worked in the Khidrat village from 2006, a time when the village's 34 beris were all non-functional. As of 2013, Khidrat has 155 beris providing drinking water to 90% of the village's population (Shehfar 2015)—the total population being 1676 individuals (Census 2011, 2015). The Sambhaav, another charitable trust, has revived several beris in the Rangarh and Netsi villages. In the Nagrani village, the villagers have built 10 beris. The introduction of beris has allowed the maintenance of resource-demanding sheep and cows in addition to low-maintenance goats and camels so as to fetch extra income. Since 2005, the Panchayati Raj Institutions (PRI) under the Mahatma Gandhi National Rural Employment Guarantee Act (MGNREGA) has constructed around 300 beris in Rajasthan. Each beri supports 8–10 families and 150–250 livestock. The scheduled caste villages were the most neglected as regards access to clean drinking water. Villagers here formed the Dalit Jagruk Samiti—affiliated to the Thar Nagrik Jagruk Manch (TJNM) in Barmer; along with the SURE (a Barmer-based NGO) a collective effort exists that intends to construct beris in these remote villages on the Rajasthan–Pakistan border (Down to Earth 2015). Beris have provided a major source of income for local artisans, who can fetch good money from each construction. Growing knowledge about beris among the citizenry has allowed greater more involved in both government and non-government initiatives centred on beri revival. This awareness and interest significantly contributed to the increasing number of beris in west Rajasthan.

5.5   Bawris/Baoris

Bawris—also known as baolis or stepwells—are an age-old method of rainwater harvesting, which has existed for at least 1000 years. These are dugout deep wells or ponds wherein the stored water is accessed by climbing down several steps built around the water body which follow a unique concept. The architectural magnificence of stepwells makes them one of the most visually spectacular water harvesting constructions. Bawris store water for long periods of time due to reduced evaporation, and water is available throughout the summer season (Burgess and Henry 1903).

   Bawris can be round, square or rectangular in shape and either magnificent or simple in structure—depending on the builder. The construction of bawris and jhalaras (rectangular stepwells) typically begins from bottom up. The area is dug into a deep pit from where ground water can be accessed and boundary walls are then risen that surround the water body. Underground water levels determine the depth of the stepwells and the number of steps that elaborate the bawri design. The bottom of the bawri is deep and narrow so as to minimize water loss through evaporation. Bawris have a vertical shaft at the centre with water and a broader mouth that surfaces into a pool, around which steps are built to access the water. Groundwater from bottom and rainwater from above are both collected in the bawri. Bawris are built where there are nearby catchments that can recharge the groundwater. They are structured like a funnel, wherein the stem is the vertical shaft containing water until the junction of the stem and the inverted cone. The steps that surround the water take the shape of said cone, with step tiers that become broader from base to the top. The steps sometimes surround all four sides; yet often they only occupy three sides, the fourth side consisting of a regular building or temple construction. The stepwells are predominantly made with sandstone and sometimes with marble (Livingston 2002).

   At the base of the bawri, small room-like structures were often constructed with elaborate designs and served as places for rest—where the temperature is usually 5° less than outside. There is also always a temple, which is a very unique traditional feature to these stepwells. Bawris, like most water conservation structures, were considered very sacred and were often dedicated to particular God or Goddess. During the pre-independence era, the prime time of bawris, temples and resting areas were built with beautiful carvings—and many were painted in bright colours with lime-based paint. Many of these structures have been built with unique imagination, featuring the architectural excellence of sandstone and marble artisans in olden times. The mild rains work through an intricate network of catchments, khadins, naadis and taankas—the subterranean seepage from which percolates into the ground, raises the water table and recharges a deep network of aquifers, thus making water available in the bawris (Mishra 1994).

The history of bawris dates back to the age of Harappa and Mohenjo-Daro roughly 4500 years ago (Saxena 2017). During this period, square bath wells were created with steps and surrounding rooms for rest purposes. 3500 years later, India had deeper stepwells with a similar architecture. Some of the bawris are much recent—about 150 years old—whereas some are over 1000 years old. Several bawris were constructed by the royals and affluent members of the princely state. The city of Jodhpur has eight stepwells, the oldest being the Mandor bawri built in 784 AD—and the Mahamandir Jhalara, which dates back to 1660 AD. The Raniji-ki bawri built by Rani Nathavati ji in 1699 AD during her son Budh Singh's rule is one of the finest examples of bawri architecture. It is a multi-storeyed structure with places for worship on each floor and finely sculpted pillars and arches. The Chand Bawri of Abhaneri near Jaipur is by far the most famous and architecturally—and mathematically—spectacular stepwell in India. It was built by the Maharaja Chand during eighth–ninth centuries AD and was dedicated to the Goddess Harshat Mata. Although it was constructed as a temple stepwell, the Chand Bawri probably helped the kingdom of Abhaneri to face periods of drought. The Jagu bawri (1465) and the Idgah bawri (1490) are among the oldest baoris of Jodhpur; the Shiv bawri (1880) is more recent. Bundi is famous for its c. 86 bawris—the deepest of which is about 46 m deep. The Panna Meena ka kund in Jaipur, the Uday Pol Toran bawri in Udaipur, the Bhatti ke bawri, the Birkha bawri, Brij bawri, Kaluram jee ki bawri and Mandore garden bawri in Jodhpur are some of the famous, old bawris in Rajasthan.

Bawris ensured a regular supply of water, mainly for religious rites and royal ceremonies. They also served the community for daily needs, mainly for bathing and washing clothes, and served as centres for social gatherings, festivals, recreational activities and other collective endeavours. The British disfavoured this community-based water supply arrangement and started to incorporate pipelines and pumps into the system. This led to clogging, drying and a gradual dereliction of bawris.

Gradual decline has caused many of the bawris to become saline and unusable because of th accumulation of dirt and debris. These structures mostly stand out as architectural water harvesting histories. Environmentalists and local people have stepped forward so as to rescue these structures. Rajesh Joshi, an environmentalist, together with a team of volunteers, has cleaned Jodhpur's Kriya ka jhalara, Sukhdev Tiwaei jhalara, Mahila Baag jhalra and Satyanarain ji ki bawadi bawris during 2015 (Somvanshi 2015) The government of Rajasthan has taken some initiative through the launch of the Mukhyamantri Jal Swavlamban Abhiyaan in 2016 so as to renovate bawris across Rajasthan. The programme was initiated with the restoration of the

Nachan Bawri between Jaipur and Ajmer. Most bawris are still in a poor state, and their repair will take much effort and time—which is probably why the building of new bawris has not been initiated. The amount of labour, time, architectural detailing and effort involved is very great indeed—and these resources could be invested in other water harvesting structures which are easier and faster to build. Jodhpur has a single twenty-first-century stepwell—the only standing testament to the usefulness of bawris even in the modern era.

6 Conclusion

The above discussion reveals the story of evolving human relationship with the environment in western Rajasthan. The western Rajasthan is a desert with significant seasonal fluctuations in temperature and scanty rainfall. Human life in the desert has been tough and challenging. However, through creativity and adoptability, humans have created a peculiar history in this desert. The most difficult environmental component in any desert is water scarcity—a great challenge indeed, for water is needed for all biological existence. The significance of this scarce environmental component has been recognized by residents and has been accorded highest priority. The methods and tools for water conservation invented by the locals in the regions are learnt from observing nature itself—and have consistently evolved and improved over time. The story of taankas, beris, khadins, bawris and naadis is as fascinating (if not more) in the context of the desert as the story of the invention of modern medicines and artificial intelligence is in the context of the Industrial Revolution. These inventions helped humans in these regions to survive in conditions of scarcity. Water scarcity also shaped their economic, cultural and political lives in many ways. The story of adaptation and innovation in water management in western Rajasthan also shows how autonomous communities using their social capital and learning from their contexts created environmentally sustainable socio-economic practices. The region also provides a story of possibilities of human adaptation to declining water availability because of changing climatic conditions. Man has to learn from his history in different contexts so as to find the possibility of adaptation and mitigation of the challenges emerging from the environmental degradation and climate change. As mentioned above, the water conservation practices developed in western Rajasthan may be one such useful lesson and indeed a model of adaptation.

References

Ahmed E (1969) Origin and geomorphology of the Thar Desert. Ann Arid Zone 8(2):171–180
Banerji S (1952) Weather factors in the creation and maintenance of the Rajputana desert. National Institute of Sciences, New Delhi, pp 153–165

Bharara L (1980) Social aspects of drought perception in arid zone of Rajasthan. Ann Arid Zone 19(1–2):154–167

Bharara L, Seeland K (1994) Indigenous knowledge and drought in the arid zone of Rajasthan: weather prediction as a means to cope with a hazardous climate. Int Asienforum 25(1–2):53–71

Burgess J, Henry C (1903) The architectural antiquities of Northern Gujarat, more especially of the districts included in the Baroda state. In: New Imperial Serie, vol XXXII, Western India. London

CAZRI (2007) CAZRI perspective plan: vision 2025. Central Arid Zone Research Institute, Jodhpur

Census (2011, 2015) Census 2011. http://www.census2011.co.in/data/village/84034-khidrat-rajasthan.html. Accessed 27 Aug 2018

CSE (2006) Director's report 2002–2006. Centre for Science and Environment (CSE), New Delhi

CSR Portal (2018) NRM Watershed Management. https://csrrajasthangov.in/project/rainwater-harvesting-groundwater-recharge.html. Accessed 22 Aug 2018

Dhir R, Singhvi A (2012) The Thar Desert and its antiquity. Curr Sci 102(7):1001–1008

Directorate District Gazetteers (1995) Rajasthan State gazetter: volume one land and people. Government of Rajasthan, Jaipur

Down to Earth (2015) Sweet Beris of Thar: Residents dig shallow to tap sweet water reserves. https://www.downtoearth.org.in/coverage/sweet-beris-of-thar-33720. Accessed 27 August 2018

Goyal R, Issac V (2009) Rainwater harvesting through taanka in hot arid zone of India. Central Arid Zone Research Institute, Jodhpur

GRAVIS (2003) Harvesting the rains in Thar. Gram Vikas Vigyan Samiti, Jodhpur, Rajasthan

GRAVIS (2018) GRAVIS (Gramin Vikas Vigyan Samiti). http://www.gravis.org.in/index.php/our-work/water-security. Accessed 21 Aug 2018

GRAVISINDIA (2017a) Food security in the Thar Desert. https://gravisindia.wordpress.com/2017/07/15/food-security-in-the-thar-desert/. Accessed 15 Aug 2018

GRAVISINDIA (2017b) Recollection of GRAVIS' food security efforts in Thar on World Food Day. https://gravisindia.wordpress.com/2017/10/17/recollection-of-gravis-food-security-efforts-in-thar-on-world-food-day/. Accessed 15 Aug 2018

Guhathakurta P, Rajeevan M (2008) Trends in the rainfall patterns over India. Int J Climatol 28:1453–1469

Hora SL (1952) The Rajputana Desert: its value in India's national economy: a general review of the symposium on the Rajputana Desert. National Institute of Sciences, New Delhi, pp 2–3

Indian Institute of Tropical Meteorology (IITM) (2016) IITM Indian regional/subdivisional Monthly Rainfall data set. ftp://www.tropmet.res.in/pub/data/rain/iitm-subdivrf.txt. Accessed 21 Aug 2018

IRC (1988) Revival of water harvesting methods in the Indian desert. Arid Lands Newsl 26:3–8

ISRO (2016) Desertification and Land Degradation Atlas of India: Based on IRS AWiFS data of 2011–13 and 2003–05. Indian Space Research Organization Department of Space, Government of India, Ahmedabad, India

Jhunjhunuwala B (2005) Traditional agricultural and water technologies of Thar. Kalpaz Publications, New Delhi

Kar A (2014) The Thar or the great Indian sand desert. In: Landscapes and landforms of India. Springer, pp 79–90

Kar A (ed) (2017) B:8 Geomorphological field guide book on Thar desert. Indian Institute of Geomorphologists, New Delhi

Kaul R (1996) Sand dune stabilization in the Thar desert of India: a synthesis. Ann Arid Zone 35(3):225–240

Khan MA, Narain P (2003) Integrated watershed management for sustainability. In: Human impact on desert environment, pp 149–157

Khosla A (1952) Opening speech: symposium on the Rajputana desert. National Institute of Sciences, New Delhi, p 14

Kolarkar A (1990) Khadin: a sound traditional method of run off farming in Indian desert. Max Mueler Bhawan Goethe Institute, Bombay

Krishnan M (1952) Geological history of Rajasthan. National Institute of Sciences, India, New Delhi, pp 22–23

Levi R, Mishori D (2015) Water, the sacred and the commons of Rajasthan: a review of Anupam Mishra's philosophy of water. Transcience 6(2):2–25

Livingston M (2002) Steps to water: the ancient stepwells of India, 1st edn. Princeton Architectural Press, New York

Ministry of Water Resources (2016) Report of the expert committee to review available information on palaeochannels. Government of India, New Delhi

Mishra A (1993) Aaj bhi Khare hai Talaab (The ponds are still relevant), 1st edn. Gandhi Peace Foundation, New Delhi

Mishra A (1994) Rajasthan ki Rajat Boondein (The radiant drops of Rajasthan), 1st edn. Gandhi Peace Foundation, New Delhi

Narain P, Kar A (2005) Drought in Western Rajasthan: impact, coping mechanisms and management strategies. Central Arid Zone Research Institute, Jodhpur

Office of the Registrar General & Census Commissioner (2011) Linguistic survey of India. Office of the Registrar General & Census Commissioner, New Delhi

Oldham C (1893) The Saraswati and lost river of the Indian desert. R Asiatic Soc 34:49–76

Oldham R (1886) On the probable change in the geography of the Punjab and its rivers. J Asiatic Soc Bengal 55:322–343

Pant G, Parthasarthy B (1981) Some aspects of an association between the southers oscillation and Indian summer monsoon. Arch Meteorol Geophys Biokilamatologie 1329:245–252

Peerzeda AR (2017) Meet India's dam-building grandmother. BBC News, 12 December

Planning Commission (1981) National committee on the development of backward areas. Government of India, New Delhi

Prasad R, Mertia M, Narain P (2004) Khadin cultivation: a traditional runoff farming system in Indian Desert needs sustainable management. J Arid Environ 58:87–96

Roy B, Sen AK (1968) Soil map of Rajasthan. Ann Arid Zone 7(1):1–14

Rudolph SH (1963) The princely states of Rajputana: ethnic, authority and structure. Indian J Polit Sci 24(1):14–32

Saxena D (2017) Water conservation: traditional rain water harvesting systems in Rajasthan. Int J Eng Trends Technol 52(2):91–98

Shaban A (2017) Is it climate change or cyclic variation? Temporality and spatiality of rainfall in India. Working paper, pp 1–21

Sharavani (2016) Water harvesting method used in Rajasthan (Khadin). Water Management. http://www.soilmanagementindia.com/water-management/water-harvesting-system/water-harvesting-method-used-in-rajasthan-khadin-water-management/16142. Accessed 23 Aug 2018

Sharma J, Bhadra B (2014) Vedic Saraswati River Network in the Late Quaternary Period from Mansarovar to Dwaraka: perceived through Satellite Remote Sensing. New Delhi, Regional Remote Sensing Centre-West, NRSC/ISRO, Department of Space, Government of India, CAZRI Campus, Jodhpur, Rajasthan

Shehfar (2015) Beris to the rescue. https://www.downtoearth.org.in/coverage/beris-to-the-rescue-40065. Accessed 28 Aug 2018

Shyampura R, Sehgal J (1995) The Soils. Soils of Rajasthan for optimising land use. National Bureau of Soil Survey and Land Use Planning, Nagpur, pp 20–31

Singh S (ed) (1977) Geomorphological investigations of Rajasthan desert. Central Arid Zone Research Institute, Jodhpur

Sinha J, Sinha MK, Adapa UR (2013) Flow—River rejuvenation in India: impact of Tarun Bharat Sangh's work. Swedish International Development Cooperation Agency, s.l.

Somvanshi A (2015) Rescuing the stepwells of Jodhpur. https://www.downtoearth.org.in/gallery/rescuing-the-stepwells-of-jodhpur-51552. Accessed 29 Aug 2018

Tamhane V (1952) Soils of Rajputana and Sind Deserts. National Institute of Sciences, New Delhi, pp 254–259

Tarun Bharat Sangh (2009) Restoring life and hope to a barren land: 25 years of evolution. Tarun Bharat Sangh, Alwar, Rajasthan

The Better India (2015) How one woman made 100 villages in Rajasthan fertile using traditional water harvesting methods. https://www.thebetterindia.com/21899/amla-ruia-check-dams-rajasthan-aakar-charitable-trust/. Accessed 18 Aug 2018

The Rajputana Gazetteer (1879) The Rajputana Gazetteer: vol I, pp 2–3

Tod J (1902) Annals and antiquities of Rajasthan or the Central and Western Rajput States of India. Coronation Edition (Volume I and Volume II) ed. The Society for the Recitation of Indian Literature, Calcutta

Tod J (1920) Sketch of the Indian Desert. Annals and antiquities of Rajasthan or the Central and Western Rajput States of India, 1st edn. Oxford University Press, Humphrey Milford, pp 1257–1275

Vangani N, Sharma K, Chaterji P (1988) Tanka—a reliable system of rainwater harvesting in the Indian Desert. Central Arid Zone Research Institute, Jodhpur

Chapter 13
Assessing Infrastructural Encroachment and Fragmentation in the East Kolkata Wetlands

Sk. Mafizul Haque

1 Introduction

Water has never received any attention as long as it was available on sufficient amounts over the land surface system. Even on non-dry lands, water can become scarce in highly populated urban areas. Various scholars and researchers have examined these types of situation; they point towards high amounts of water demand, increasing diversion of land resources towards further development projects and consequent less permeable surfaces, etc., as the main factors leading to water poverty. Metropolitan expansion over the deltaic plain has led to an enormous encroachment upon water bodies—including the ecologically sensitive wetlands. Starting in the last decade of the twentieth century, immigration towards the leading metro centres of developing nations has increased such infringements at an alarming rate. The formation of settlements around water sources is a practice of human civilization since time immemorial. In the present conjuncture, the process of urbanization under the influence of the market economy initiated in the post-liberalization era has made infrastructural development a rapid force to reckon with (Shaw 2015). Consequently, the functional identity of existing natural resources has changed drastically—particularly at the immediate surroundings of the growing urban centres. Urban development is not possible without a human-induced modification of land cover and use; this can result in a fragmentation of the landscape and large-scale environmental degradation within a short period of time (Grimm 2008; UN HABITAT 2003). In earlier times, isolated and small-scale practices of such conversions created the least impact on natural systems and indeed benefited from a number of symbiosis with natural systems. By contrast, alteration of land use and land cover has great environmental consequences; the terms used to describe this interaction between urban and natural systems include ecosystem process, biogeochemical cycle, biodiversity and growing human activities (Kilic et al. 2006; Xiao et al. 2006; Aguilar and Ward 2003). Land

Sk. Mafizul Haque (✉)
Department of Geography, University of Calcutta, 35, B. C. Road, Kolkata 700019, India

© Springer Nature Switzerland AG 2020
S. Bandyopadhyay et al. (eds.), *Water Management in South Asia*,
Contemporary South Asian Studies, https://doi.org/10.1007/978-3-030-35237-0_13

modification by progressive urban expansion and encroachment into the vast rural area affects both local and regional ecologies (McDonnell et al. 1997; Baker et al. 2001; Pabi 2007) in different corners of the world. Indeed, ecological costs are local, regional and global (Cosentino and Schooley 2018; Mills 2007), as are changes in water environments and their behaviour under the influence of this new land–society–atmosphere interaction. The Ramsar Convention declared the science of 'water environment' to include not only bodies of water themselves, but also (a) extended banks of water bodies, (b) areas which experience seasonal inundation, (c) lands surrounding flooded pockets, (d) feeder areas of 'pond ecology', (e) wastewater-based agriculture land and (f) areas which support wetland ecosystems. However, the pace of urbanization reduces the availability of arable lands within the urban territory (Dewan and Yamaguchi 2009b); it also easily modifies and destroys both flora and faunas' habitats (Alphan 2003)—so being that species extinctions can by themselves much reduce an ecosystem's net primary production (Nagendra et al. 2003). Further consequences of these impacts include a threat on food security (Brown 1995), the shattering of landscape unity and increased patch fragmentation (Dewan and Yamaguchi 2009a; Grimm 2008; Herold et al. 2002). Finally, all these modifications have negative impacts on human societies (Dewan et al. 2010; Nagendra et al. 2003; Liu et al. 2002).

Water-centric riverine urban growth has constituted a traditional imprint of Indian urbanization since time immemorial. Kolkata, a medieval urban centre, has that opportunity to enjoy vast water resources due to its position in the deltaic region of the Ganga River system. The journey of Sutanuti, Gobindapur and Kolkata from a cluster of rural settlements to one of the Asia's largest metropolises, is influenced considerably by the river Hooghly and its ecological relationship with settlements. Kolkata's urban identity has always been—directly or indirectly—controlled by this river system. While Kolkata is bordered by the riverfront in the west and bounded by tropical water environments in the east, it also contained a number of ponds within the municipal limit. Very few urban centres in the world are as substantially endowed with rivers, ponds and wetlands as Kolkata and its surroundings. However, the uncontrolled expansion of urban environments has created a threat to both surface and groundwater resources. Hartshorn (1992) has mentioned that land fragmentation and alteration of its settings mostly feed urban needs—yet it also generates multiple effects on environmental processes. Population influx and physical expansion of urban built-up around the metropolitan cores are considered critical factors in issues regarding land and its fragmentation—and this is the case in both developed and developing countries. Bose (2008) stated that Kolkata's unexpected developmental activities have surged since 1996 and continue to this day. Owing to the high rate of population concentration within the territorial limit of the premature metropolis, Kolkata has remained over-bounded for a long time. In the last four decades, the fringe area of Kolkata has absorbed more population than its downtown, as the city has experienced a rapid outward extension of housing infrastructures, functional agglomerations and structural modifications (Pathak 2011; Haque et al. 2019). The fragmentation of natural land segments in the East Kolkata Wetlands (EKW) poses immense threats to the ecological function of tropical wetlands both within and in the immediate surroundings of the urban limits (Dey and Banerjee 2013a). Urban developments much upset fragile ecological relations—and ultimately encroached on the entire ecosystem (Haque and Bandyopadhyay 2012; Shaw 2015).

At the doorstep of these worldwide unique wetlands, the warrior has surrendered—all the urges from both environmentalists and literary scholars have failed before the neoliberal activities under the patronage of 'advanced' economies. No urban development paid homage to the glory of this unique city identity—all instead capitalized on this water environment strictly so as to promote real estate activities. Sometimes advertisements are paradoxically seen selling plots under the EKW jurisdiction itself. In many places, these types of water bodies in the city proper have no value to the common people—many indeed treat these places only as breeding chambers for mosquitoes in a dire need of reinvention. It is too difficult to let the citizenry know the actual benefits of a wetland's function and what they really mean for human survival—most importantly, what could be the consequences of their loss. The eastern fringe of the EKW in 2017 'was not just filling up of the water bodies but yet numbers set of buildings coming up' (Ghosh 2017). The main objective of this article is to scrutinize the nature and trends of infrastructural activities from 1984 using a modern earth observatory platform and tools under the arena of a geographic information system. This article also depicts how infrastructural development came to threaten the wetland ecosystem. 1984 was chosen as the benchmark for the study because the Kolkata Municipal Corporation added three municipalities within its old that year—among which is the Jadavpur Municipality, which is located at the western fringe of EKW. Various empirical experiences and mapping outputs—along with the existing literature—have been incorporated in the final assessment regarding the fragmentation of the world's largest and unique wastewater treatment practices in Ramsar-enlisted wetlands. As a wetland of international importance, the EKW is a natural resource wherein a symbiotic ecosystem sutures science and social justice—apart from providing goods and services.

2   Data Set and Methodology

The data from different fields of geospatial and earth observatory technology (satellites and Google view)—along with the empirical observations—have been used so as to understand the relationship between man and environment in the study area. Datasheet details are furnished in Table 13.1.

For the change detection, the entire EKW area was divided into thirty-six equal segments with the help of 10° interval axis forming radials from a geo-centre—wherein the perpendicular lines are considered as geo-centre lines on the basis of X-axis and Y-axis. This geo-centre is located near the Bamanghata Bridge—coordinated at 28°27′56″ E and 22°31′18″ N. Here, the boundary of the EKW for the year 1984 has been drawn based on the same visualizing characteristics of the EKW and considering seven parameters: tone, textures, size, shape, pattern, shadow and site-situation. These were also perceived in 2002 satellite images and Google views.

Table 13.1 Catalogue of different types of data sources and its uses

Data type	Sub-type		Nature	Source	Purpose
Earth observatory images	Sensor	MSS	$r = 149, p = 044$ January 1984	https:// earthexplorer. usgs.gov	Image classification For LULC, waterbody mapping
		TM	$r = 138, p = 044$ January 1988, 1994; December 2010, 2005		
		OLI-1	$r = 138, p = 044$ November 2018		
	Aerial		Aerial view	Google Earth 1984, 1994, 2005, 2017	Visualization of infrastructures and LULC changed detection
Maps	Topographical sheets		79 B/6 on 1:50,000 79 B/5 on 1:50,000 (surveyed on 1958–'59) Toposheet on inch map (surveyed on 1922–'24)	SOI—1973 SOI—1968	Base map for assessing the previous land resources and infrastructures
	Guide map		Open street map	openstreetmap.org 2017	Base map and preparation of the topology of road network
	Other thematic maps		Growth of Calcutta City Urban land use—Calcutta City on 1:25,000 base map of EKW	CMPO—1963 NATMO—1998 EKWMA—2010	LULC map and preparation of EKW base map

Source Prepared by the author

For this purpose, the SOI toposheets on 1:63,360 scale (inch map), a waste recycling region map prepared by Ghosh in 1987, have been considered for all remotely sensed data outputs. The ten-point Likert scale has examined the qualitative nature of different features through the spatial data mining process. In this data extraction method, point-based (e.g. trees), line-based (e.g. roads, canal, channels, etc.) and polygon-based (e.g. open lands, water bodies, agricultural lands, etc.) features have been considered along the radial axis. Finally, the rate of linear extension of the built-up environment/shrink of water environment has been calculated by using the following formula:

$$D(c) = \sum_{i=1}^{n=max} (1 + \sum f(TT'PSA)t_1) \sim (1 + \sum f(TT'PSA)t_2)/t_1 \sim t_2$$

where

$D(c)$	= change in axis distance
l	= length of the axis
T	= tone of features lie between intersecting points
T'	= texture variation of features within intersecting points
P	= pattern of the features in intersecting points
S	= size of the intersecting features
A	= association of the intersecting features
t_1	= preceeding year
t_2	= succeeding year

After obtaining the changed rate, the classification output of satellite images has been overlaid on topographical maps for the purpose of understanding land utilization dynamics and infrastructural developments. For a better assessment of water bodies (which include 'bheries' or waste treatment-based fishing grounds and 'ponds' mainly used by local residents), the environment surrounding the EKW was also incorporated because infrastructures are not isolated systems but carry multiplier effects in the shape of functional agglomeration of urban activities, which gradually replace the rurality.

Evolution and Functions of the East Kolkata Wetlands (EKW):

Lowland areas in tropical monsoon are subject to seasonal inundation. During the period of high intensity of rainfall in late summer, lowland areas are filled with water and act as an inter-distributary marshy land. The East Kolkata Wetland (EKW) is such a type of brackish water, dominated by marshy land formed in the mature delta of the Ganga River system (Chattopadhyaya 1990). According to Bagchi (1944), it is assumed that owing to the loss of annual deposition of silt, the inter-tributary areas of the Rivers Hooghly and Kulti became depressed and were simultaneously enclosed by the raised tracts of the natural levee during the last phase of delta-building stage. Though the formation has largely been influenced by the Ganga River system, the tidal deposition process and the rhythm of seasonal inundation continue in and around the EKW. More specifically, this area was mainly nourished by the River Bidyadhari so as to develop the backwater swamp and spill reservoir area of the tidal bore (Chattopadhyaya 1990; Ghosh and Dutta 2019). Before the setting of the colonial foot in 1690 at Kolkata (formerly Calcutta), the hydrological regime of the study area was largely influenced by the Rivers Kulti, Bidyadhari and Adi Ganga—along with three main tidal creeks. With the development of human settlements on the Hooghly levee (in its right bank), almost all the spill channels were reclaimed and became chocked (Ghosh 1987, 1999). From the documents pertaining to the Calcutta Municipal Gazette (1964) regarding the 'Reclamation of Salt Lake', it is found that the wetland was on vast area stretched up to Dum Dum and even in close proximity to the River Hooghly. In that time, the entire wetland was dissected into two parts by a sediment ridge and connected by a channel. During the first half of the British era, this area was abandoned due to the adverse conditions for sustaining a human settlement. After the frequent occurrences of malaria, plague and other diseases in the city proper, Kolkata's sewage system was connected with the EKW so as to discharge the waste there instead of through the River Hooghly (EKWMA 2011). The drainage scheme of Kolkata constructed canals, sluices and bridges across this wetland system—and the magnitude of interventions was increased with the progress of various project schemes in the last two decades of nineteenth century (Mukerjee 1938; DEC 1945; Ghosh 1999). That was the first attempt to fragment the entire wetland into different segments. In the year 1865, an unwanted act of filling up the 2.59 km^2 of wetland area for the dumping of garbage and sewage farming was undertaken. This was the first effort towards the modification of the tropical wetland, i.e. the EKW. Thereafter, multiple instances of human intervention have become primary modes of wetland modification and re-fragmentation. Until the 1980s, a range of developmental projects and activities were introduced at the fringe of this area, which further fragmented and altered the wetland system beyond proportion. Therefore, various imprints of human intervention are to be considered factors in landscape change and land use modification in the EKW.

Based on Costanza et al. (1997), the estimated world's ecosystem services found that wetlands are almost 75% more valuable than lakes and river systems, 64 times more than grasslands and rangelands and 15 times more than forests. This is much more relevant still in the present day's fragile scenario; the EKW's functions are manifold and extremely significant for metropolitan survival. Before human intervention, this area was used for navigation through selected channels and trade routes (Ghosh 1987, Ghosh and Dutta 2019). But the situation has changed dramatically. For its spatio-functional dimension, the EKW is now one of the world's largest and oldest resource recovery land units—presenting all prosperities of agriculture activities, including aquaculture and horticulture. Urban agriculture is crucial because of the economic opportunities and venues for social justice it provides. In the report (Perspective Plan for Calcutta: 2011) of the West Bengal State Planning Board (1990), it is clearly mentioned that the EKW is required for flood control, regulation of water quality, treatment of wastewater, recharging the groundwater, pollution abatement and practices of pisciculture. All these activities need to be entirely guided by local people through their indigenous methods of resource utilization. In the last three decades, a vast portion of fisheries (locally known as 'bheries') in the south and south-eastern part of the EKW has been converted for agricultural purposes. Due to the presence of high intensity as well as vertical descend of solar energy, this wetland area runs a strong and unique ecosystem intended to reduce the NH4-N, COD, BOD, phenol, Cb and Pb types of pollutants naturally and without any economic cost. Due to the presence of huge nutrient diversity as well as essential metals, this ecosystem easily supports up to tertiary producers (Bhattacharya et al. 2012). Besides, the EKW is considered as the repository of millions of flora and fauna species and has a greater role as a carbon sink and its sequestration.

For the above-mentioned manifold and multifunctional activities along with its unique ecosystem support, the EKW's 12,500 ha situated near Kolkata has been listed in 2002 in the Ramsar sites as a wetland of international importance. In spite of being a Ramsar site, this area has, unfortunately, become a playground for myriads of developmental activities. Now, the EKW is facing several difficulties to sustain its ecological values—functionally as well as spatially. For management convenience, the authority EKWMA has divided (under the Wetland Rules 2010) the whole wetland area into two categories: (i) core wetland area—which includes the 'bheries' used for sewage-fed fisheries and productive farming lands, and (ii) non-core wetland area, which comprises only paddy fields and settlement areas. Presently, this waste recycling region produces 1500 tons of fish and 3600 tons of vegetable per year and has sheltered nearly 61,000 inhabitants in its 125 km^2 area. Out of its total geographical limits, fisheries are practised in 58 km^2 and agriculture over a 47 km^2 area—along with the purification of 30,000 tons of sewage per day through this natural system.

Present Scenarios of Infrastructural Advancement in the EKW:

Residential buildings and their ancillary structural elements comprise 'infrastructures' in a built-up environment of an urban setting. In this article, all kinds of land utilization practices other than the previously mentioned 'water environment' are treated as infrastructure. From the SOI toposheet surveyed in 1922–'24 and 1958–'59, it is found that the entire EKW was divided into a number of permanent water bodies meandered by earthen embankments in this rare example of wetland practices (Fig. 13.1). In 1969, the redistribution of land under land reforms led to the fill-up of approximately 25 km^2 of water bodies for the purpose of conversion into paddy fields. Gradually, the Eastern Metropolitan (EM) Bypass and the Salt Lake (Bidhannagar Municipality) City were constructed on reclaimed wetlands; this opened up the core wetland area and made it more accessible and attractive to real estate speculators. From the report of a project carried out by the University of Calcutta and PAN network, it is found that between 2005 and 2011 nearly 10% of the land's area has been converted to urban purposes (Dey and Banerjee 2013b). Apart from waste dumping and its treatment-related structural developments, the detailed account (Table 13.2) on various types of infrastructural surges (from 1984 to 2017) within the limits of the EKW is described below:

(1) Metalled Roads: A numbers of metal roads have been extended (Fig. 13.3) for several hundred kilometres through this eco-sensitive wetland area. Five have played a major role in dissecting the EKW. (i) The Basanti Highway—owing to its two-lane expressway service and state highway status, this road is popular for approaching the Sunderban region. Although it was constructed on the northern embankment of the storm water outlet canal, it dissected water bodies near the Bantala sluice gate area where it merged with the southern bank of the Central Lake canal. It also connects with numerous feeder roads on both sides.

(a)   **(b)**

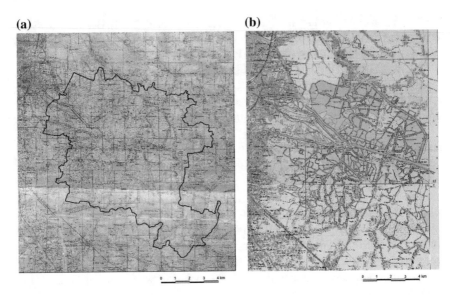

Fig. 13.1 EKW and its surrounding environments: **a** Survey of India (SOI) toposheet surveyed in 1922–1924, **b** part of wetlands in Survey of India (SOI) toposheets surveyed in 1958–1959. *Source* Prepared by the author based on SOI toposheet

(ii) The Nazirabad Road was extended eastwards from the Anandapur Bus Stand and connected with the Sonarpur-Bamanghata Road as the Bajbarantala Road. Situated at the heart of the EKW, this road also connects with fourteen medium- and short-distanced metal roads. (iii) The Tardah-Kheyadah Road— actually an extended pathway of the Nazirabad Road which was extended westwards and joined the Bhajendrahat Road. This road connects further with six other metal roads. There are a lesser number of settlements linked to this road for direct services. It is mainly built on the right embankment of the Bainala Khal. (iv) The Sonarpur-Bantala Road, which extents in a north–south direction (connecting the Sonarpur and Basanti Expressways) and divides the south EKW into two parts. It also feeds numerous other metal roads in both sides; almost all adjacent land segments are used for either settlement or agricultural purposes. (v) The Sree Nagar Main Road creates a mega push factor for all types of developmental activities by filling-up water bodies in the Mukundapur area under the south-western section of the EKW. A complete road network and its high-intensity topology have been formulated based on this road's extended service areas.

Besides these roads, over a hundred short-distance metal roads have been constructed—or are in ongoing process. As a result, almost all sections of the EKW are now easy to access by the motorcar. From the map (Fig. 13.3), it is observed that a well-planned road network has successfully developed in the Chingrighata and Mukundapur areas. These types of road connectivity are also observed in the other parts of the south-western and southern segments of the

Table 13.2 Major developmental activities initiated in the limits of the EKW, 1984–2017

Land segments	Axis from the geo-centre	Major developmental activities initiated		
		1984–1994	1994–2005	2005–2017
North-east EKW	10	Non-metal road	Landfilling and fragmentation, metal road	Fragmentation, non-metal road
	20	Landfilling	Fragmentation	Non-metal road, building
	30	Trees	Trees, non-metal road, agri. land	Agri. land, building, non-metal road
	40	Trees	Land fragmentation, agri. land	The building complex, open land
	50	Trees, land fragmentation	Agri. land, open land	Land fragmentation and filling
	60	Dry and open land	Trees, land fragmentation, non-metal road, agri. land	Land fragmentation, metal road, building
	70	Land fragmentation, trees	Agri. land, trees, non-metal road	The building, land fragmentation
	80	Trees, land fragmentation	Agri. land, land fragmentation	Agri. land, building, land fragmentation
	90	Land fragmentation, trees	Land fragmentation, agri. land, trees	Trees, agri. land, building, land fragmentation
South-east EKW	100	Trees	Agri. land, land fragmentation, building, non-metal road	Metal road, land fragmentation, agri. land, building
	110	Trees, bushes	Land fragmentation, metal road, agri. land	Metal road, canal, building complex, high rises
	120	Bushes	Trees, land fragmentation, metal road, building	Canal, metal road, high rises
	130	Bushes	Trees, land fragmentation	Agri. land, building, fisheries
	140	Open land, trees	Trees, bushes, agri. land	Agri. land, fisheries, building

(continued)

Table 13.2 (continued)

Land segments	Axis from the geo-centre	Major developmental activities initiated		
		1984–1994	1994–2005	2005–2017
	150	Non-metal road	Agri. land, building	Agri. land, fisheries, building
	160	Open land, non-metal road	Canal, landfilling and fragmentation	The building, metal road, agri. land, fisheries, land fragmentation
	170	Open land, landfilling	Non-metal road, building	Metal road, non-metal road, landfilling, agri. land, building, fisheries
	180	Landfilling, agri. land, building	Non-metal road, agri. land, building	Agri. land, building, metal road
South-west EKW	190	Land fragmentation, building, non-metal road	Metal road, building, agri. land	Land fragmentation, metal road, building
	200	Land fragmentation and filling, fisheries	Canal, metal road, building	Metal road, building, fisheries, agri. land
	210	Trees, bushes, land fragmentation, building	Agri. land, building, metal road, land fragmentation, fisheries	The building, metal road, land fragmentation, high rises
	220	Dry and open land	Non-metal road, land fragmentation, building, high rises	Land fragmentation, building, high rises
	230	Non-metal road, land fragmentation	Land fragmentation, fisheries, building, high rises	Land fragmentation, metal road, building, high rises, fisheries
	240	Metal road, landfilling, trees	Metal road, land fragmentation, building	Metal road, building, high rises
	250	Trees, bushes	Metal road, building, land fragmentation, high rises	Land fragmentation, building, metal road, high rises

(continued)

Table 13.2 (continued)

Land segments	Axis from the geo-centre	Major developmental activities initiated		
		1984–1994	1994–2005	2005–2017
	260	Metal road, landfilling and fragmentation, building	Land fragmentation, building, high rises, metal road	The building, land fragmentation, agri. land
	270	Landfilling and fragmentation, metal road	Land fragmentation, agri. land, fisheries, building, high rises	Land fragmentation, building, metal road, high rises
North-west EKW	280	Metal road, landfilling and fragmentation	Land fragmentation, fisheries	Fisheries
	290	Dig the land	High rises, metal road, landfilling and fragmentation	Agri. land, metal road, building, high rises
	300	Metal road, landfilling	Agri. road, metal road, land fragmentation	Land fragmentation, agri. land, metal road
	310	Building, metal road	Land fragmentation	
	320	Metal road, land fragmentation	Solid waste dumping site	Landfilling and fragmentation, fisheries
	330	Land fragmentation	The solid waste dumping site	The solid waste dumping site
	340	Landfilling and fragmentation, metal road	Land fragmentation	The building, high rises, land fragmentation
	350	Landfilling and fragmentation, metal road	Road, land fragmentation, building	High rises, building complex
	360	Landfilling and fragmentation	Metal road, building, land fragmentation	Land fragmentation, building, metal road, high rises

Source Computed by the author based on SOI toposheets, Google Earth and LANDSAT-TM, OLI-1
The axis is drawn from the geo-centre (based on 1984 spatial extension of wetlands) located at Bamanghata Bridge, 28°27′56″ E; 22°31′18″ N

EKW as well as in fringe areas lying in the north-eastern side. The crucial fact here underlined is that the EKW is now bordered by various important metal roads and expressways, which has catered to the need to agglomerate all kinds of urban infrastructural functions. The EM Bypass in the western periphery constitutes a good example. While only 370 m of this expressway is located between two patches of the Captain Bheri under the EKW, it hosts a great role in the fragmentation of the water ecosystem in the EKW's western fringe. The upcoming metro project will also amplify developmental opportunities in almost all the peripheral regions of this Ramsar site. The Major Arterial Road (east–west) lies in the north and north-western sides of the EKW and is now providing different types of infrastructural activities under the rhetoric of the Newtown project. The Ganga Joara Road in the southern fringe of the EKW is gradually becoming more lucrative for both residential and retailing activities.

(2) Unmetalled Roads: The huge number of unmetalled roads, though short in length and very narrow by comparison with metalled structures, has led to a direct fragmentation of the 'bheries'. While fundamentally they are earthen in composition, at some places burnt bricks are also used so as to cover the surface of these roads, whose main function is to connect wooded settlements. Some of these are covered with bitumen and bricks, as well as basalt fragments, especially in compact, wooded settlement areas. These roads are responsible for creating a network topology in the areas between the main roads and the interior parts of the wetland system. The newly occupied areas of Hatgacha, Kulberis, Hadia, Beonta, Garal, Tarda, Nayabad, Tihuria, etc., are very popular among the citizenry because of this type of road formulation. They answer the eagerness to reach waterbodies—and gnawn away the segments part by part.

(3) Lock gates: Several lock gates were built in the channels and 'bheries' so as to regulate the flow of water under a controlled system. They were basically earthen and wooded structures in earlier days, but now they present a concrete structure with regulated instruments. In the case of 'bheries', it has been used for both inlet and outlet purposes so as to control and regulate the sewage intake capacity.

(4) Towers: Two high-voltage power lines are extended through this area from west–east direction. One is almost parallel to the Basanti Highway and another is located in the north-western portion of the EKW. From a report by the ERLDC, it is found that another four transmission lines of 220 and 400 kV were penetrated in a south–north direction through the sub-surface layer. Apart from power line towers, numerous towers were also set up for mobile telecommunication facilities. These are commonly seen in the north, west and south-west portion.

(5) Bridge and Culverts: Numerous small-scale bridge and culverts were constructed over the main channels and on the water-drained paths. Besides the concrete structures, earthen and wooded bridges are also noticed in and around almost all the 'bheries' under the EKW. However, concrete and permanent structures are most vital for maximizing the accessibility to the interior of the water bodies. They facilitate the process of raw material transportation for

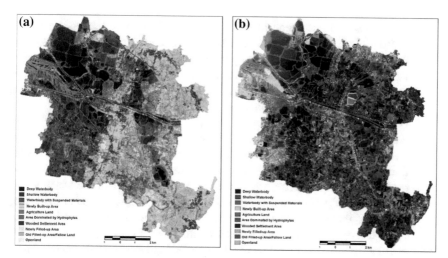

Fig. 13.2 Different types of existing LULC in Standard FCC output of **a** TM sensor, March 2000; **b** OLI-1 sensor, November 2018 of Landsat satellite. *Source* Prepared by the author based on Landsat, TM (2000), OLI-1 (2018), https://www.usgs.gov.org

the construction of other infrastructures in the filled-up waterbody segments. These also are used so as to regulate the direction of wastewater flow. In some places, real estate agencies built these types of structures for the purpose of promoting land resources as modes of consumption. Gradually, these structures link all metalled and unmetalled roads and create pathways for approaching the new and unexplored areas of waterbodies. Three large segments of land—i.e. Nalban, Nicco Park and Aquatica located in the north fringe of EKW—have been used as popular water sports activity centres for entertainment purposes.

(6)   Residential Pockets: Earlier isolated farmhouse areas (Fig. 13.2) are now becoming residential pockets with migrant populations—most of whom have very little connection with fishing activities. At present, residents, reluctant to accept the traditional job options related to waste recycling processes, are conspicuously shifting from a farming to a non-farming economy while welcoming the urban setting in the wetland environment. These residential areas are rural in nature and are surrounded by wooded trees. Road site linkage and strong canopy cover are easily assessed in the earth observatory outputs. In the present study, these types of land utilization are treated as wooded settlements. The water bodies surrounding these wooded settlement areas are largely fragmented as well, very small in size and only used for local domestic purposes. These areas are also characterized by the presence of wood patches, and most of the buildings are single-storied concrete houses; semi-pucca and permanent houses have also developed in these types of residential pockets. Although two- and three-storied buildings are few in number, they continuously increase in area and capture the new open land, eventually reaching the surrounding small waterbodies. Urban facilities are continuously increasing

within these wooded rural identities, particularly in the filled land segments of the EKW in its central, north-eastern and south-eastern portions. Out of these newly emerged residential pockets, a few are small in size and located on the small areas between two waterbodies. A small number of citizens live in these isolated residential pockets between two waterbodies and engage in water environment-related activities and waste treatment works.

(7) High-rise Residentials: The most emerging issue regarding the EKW is the erection of high-rise (more than four-storied) buildings for residential purposes. Numerous high-rise buildings have been constructed in scattered positions through the advancement of real estate activities. Besides, the development of transport infrastructures accelerated activities in the western and south-western portion of the EKW. From the SOI maps and other evidence, it is found that various types of land segments were converted into high rises in the last decade. The Atghara, Ranabhutia and Mukundapur areas are most popular for the transformation of farm cum orchards and open lands into high-rise condominiums. The high-rise developmental scenario is now prevailing in Paschim Chowbaga, Khatipota, Nayabadh and Bhaganbanpur through the direct alteration of waterbodies. In the east, Bantala and Hadia mouza also experience the same scenario. In most cases, these types of residential areas create a managed land segmentation, whereby they capitalize the previous water bodies by creating swimming pools or concretize them as water chambers. The buildings, like gated communities, cater to the high-end consumer segment. With the advancement of road networks, the EKW's central portion is now becoming more suitable for direct filling up as well as land fragmentation and transformation of waterbodies into different land segments.

(8) High-rise Commercials: Areas in and around Mukundapur, Choubhaga, Ranabhutia, Jagatipota, Atghara and Karimpur mouza have all attracted commercial developments. Due to close proximity and functional agglomeration along the EM Bypass, different types of institutional set-ups, commercial hubs, retail shops, healthcare institutions, academic centres and entrepreneurial activities have spawned up in these localities. These have all acted as trigger engines for mutating land utilization practices. Functional activities for local services are more acceptable in the open land, which is to convert and replace the small patches of waterbodies. In the central portion of the EKW, the mixed use of high-rise building—i.e. residential cum commercial—is also emerging at an alarming rate. The complete association between high-rise residential and commercial hubs creates a new built-up enclave with a new social identity. Since residents in these areas are mostly outsiders, they have no psychological attachment to this unique wetland system.

(9) Concretization of Banks: In the west, north and south-western areas, the older channels are fully controlled by concretized banks and in fact converted into drains. In some places, the channel sides are fenced with wooded fencings so as to guide flow direction and avoid inundation during the monsoon season. The small water patches are also experiencing a similar concretization and becoming de facto water chambers with concretized banks. All water bodies within

the built environment are totally detached from any types of either inlet or outlet connections. The water bodies in the EKW's fringe areas have controlled bank systems and beautification imprints. Yet large 'bheries'—under both government and private agency-operated management systems—have earthen banks with linear stretches of coconut trees. Sometimes free-floating hydrophytes are also grown in the 'bheries' for increased production. These banks provide a strong ecological support to the bio-degradation of liquid waste for the sustenance of the EKW. Within a few years, the water from large 'bheries' has been drained out completely and dried up; the sludge is dug and the bank is reconstructed with sludge materials. Gradually, these banks become strong chambers of primary components of the 'pond ecosystem'.

(10) Societal Assets: The most sensitive and important practices in land conversion processes within the EKW concern the installation of social assets. In this study, local clubs, small temples, street corners with religious idols, common collection points of purified drinking water, etc., are treated as social assets. They represent an important signature of land commercialization near or in the 'bheries'. These types of practices are very common in the central and south-eastern sides of the EKW. Just after or just before the filling of fragmented and small waterbodies, these social assets are strategically supplanted in any portion of the land segment, creating an image of a social milieu—a place ostensibly infused with sentiment. After a few years, the land is handed over to real estate developers and different infrastructures for class-based consumption are built. Owing to retail and commercial activities associated with real estate development, the local citizenry tends to welcome these practices. In the last five years, real estate activities have become rampant with the installation of tin fencing, under the control of local leaders and goons.

Along with these unnoticed practices within the limits of the EKW, the immediate surroundings (fringe areas) also experience a piling up of built-up and infrastructural formations (Fig. 13.3). However, the direction, nature and characteristics of these infrastructure developments are highly varied

(1) The north-western fringe: This surrounding is entirely fenced by expressways and characterized by a compact built-up environment. The interface area lies under the KMC and Bidhannagar Municipal Corporation jurisdiction. A proposed metro rail line has also been designed in this fringe area. Since 1992, the transport topology along the EM Bypass has provided innumerable developmental options and accelerated the multiplier effects for infrastructural development in the EKW. Two world-class hotels, three large-scale built enclaves and a number of high rises along with compact residential structures have been developed recently in the immediate western surrounding of the EKW. In the last three decades, a commercial IT hub with numerous high rises has been initiated in the Sector V and Mahisbathan area.

(2) North-east fringe: This area has a combination of mixed land use covering open lands, agricultural land, orchards, wooded settlements and scattered high-rise

Fig. 13.3 Development and encroachment of road network into the EKW, 2015. *Source* Prepared by the author

apartments. Areas adjoining the Newtown have been fully converted to various development projects—including residential and commercial activities—whereas the eastern sites are mainly characterized by rurality. Yet the proliferation of Major Arterial Roads has created ample opportunities for future infrastructure development. The most critical situation is that the previous agricultural lands have been converted into residential units at an alarming rate.

(3) The south-eastern fringe: Vastly rural and in the presence of orchards, this fringe area is mostly free from large-scale built-up development; yet due to transport infrastructure initiation in the area, lots of potential for developmental projects are likely to materialize. Presently, compact wooded settlements, agricultural land, small patches of water bodies and ponds and segmented land plots are the main scenario of land utilization practices. Because of the presence of high connectivity of railways and the Canning Road, various types of infrastructural development activities have already commenced in the south.

(4) The south-western fringe: Presently, this surrounding is treated as a hotspot for different kinds of infrastructural developments. The new functional node of the metro city Kolkata is operating since 1990 in this fringe area; recently, it has engulfed the water environment of the EKW. Six international hospitals and healthcare centres, four international schools, several management

institutions, a few government offices, large-scale high-rise enclaves, corporate offices, godowns and so on have been developed in this fringe area in the last two decades. The west fringe of the EKW stretching from Mukundapur to Panchannagram has bounded the city's most popular high rises.

These overlaid, variegated and differentiated infrastructural components pose a serious threat to the world's most prestigious Ramsar sites—resulting in fragmentation, land conversion and an extension of the built-up horizon. Ultimately, an ecological crisis is unleashed.

3 Fragmentation of the East Kolkata Wetlands

The recent fragmentation scenario of the EKW is most crucial (Fig. 13.5). As all national parks, the Ramsar site should be treated as a matter of state policy orientation; any construction should be demolished immediately (Ghosh 2005). In its draft proposal (2011), the EKWMA has considered that the core wetland area will be controlled by central government rules, whereas the non-core area will remain under state authority. As the eastwards of the KMC are presently experiencing more population pressure, a large number of areas from the water environment are being devoured by built-up structures (Shaw 2015; Pathak 2011). Under the growing scenario of a market economy in the post-liberalization era, the interface between development activities and the natural ecosystem becomes fragile and porous. On the face of it, government initiatives remain of paper—while, paradoxically, continuing mostly unavailable to the public. Securing the EKW from developmental tragedy is a daunting challenge. In the last thirty-five years, this tropical wetland system has been gradually fragmented in a myriad way and through various activities. The temporal variation of its fragmentation has been sketched in the following diagram (Fig. 13.4).

Based on the four perpendicular axes of geo-centre, the entire area has been divided into four parts (Figs. 13.4 and 13.5) for this assessment: north-west (NW), north-east (NE), south-east (SE) and south-west (SW).

The water environment located in the north-western portion of the EKW is in comparatively better condition, and a strong sewage treatment system is operated along with profitable aquaculture practices. In the 1980s, natural waterbodies were partially filled up for rehabilitation programmes targeting a large number of dislocated people from the western part of Bidhannagar. Gradually, different kinds of infrastructure for the development of local lowland areas have been constructed and concretization became a regular phenomenon. Yet in the first decade of the twentieth century, restructuring of old buildings and the construction of new high rises have been much undertaken in the Naobhanga-Chingrighatta area. Due to the close proximity to IT hubs in the Sector V and Newtown areas, this locality is now highly attractive to citizens working in the IT sector. Due to the propagation of metalled roads, the fragmentation of waterbodies has been noticed along the northern side of the Belaghata-Naobhanga canal. Except for this mixed residential area, a large

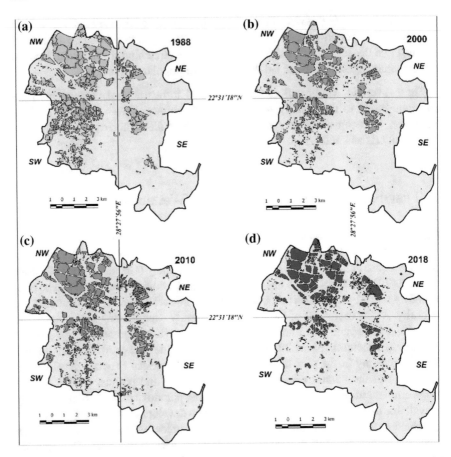

Fig. 13.4 Temporal variation of the 'bheries' in EKW and nature of its fragmentation from 1988 to 2018

part of the EKW has been maintained by various government and private agencies through strong practices of aquaculture using diversification and bifurcation of water flow direction. During field observation in 2017, several 'bheries' were found in a dry condition—with excavation activities undertaken under the Benfish authority. The linear extension (Fig. 13.6) of the built-up environment in this section is very low—about 0.034 km/year. As this portion is largely part of the core wetland area, the rate of direct fragmentation is almost negligible. On the other hand, land use conversion and fragmentation in the north-eastern segment of the EKW are quite different. The eastern periphery of this segment has seasonally dried up, and there is a mixed assembly of different-sized waterbodies. Therefore, temporary practices of agricultural activities have regularly occurred with the extension of non-metalled roads and rural houses. As a result, small waterbodies are being filled up. Presently, a few scattered high rises are being initiated due to the construction of a new expressway and other metalled roads—and a huge number of single-storied buildings are being

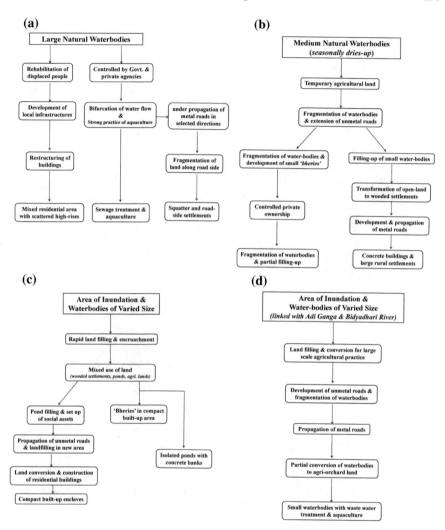

Fig. 13.5 Process of land fragmentation in EKW under: **a** north-western, **b** north-eastern, **c** south-western, **d** south-eastern segment. *Source* Prepared by author based on field observation and earth observatory information

constructed with the facilities of a large rural settlement. Waterbodies located in the western portion of the Kestopur canal have been re-fragmented and converted into small waterbodies, particularly in the Hadia mouza. A few 'bheries' in the northern portion of the Hadia are privately owned; aquaculture is practised along with recreational activities. In this section, the linear shrinkage of the water environment (i.e. extension of the built-up environment) is moderate in nature—its average rate is 0.56 km/year.

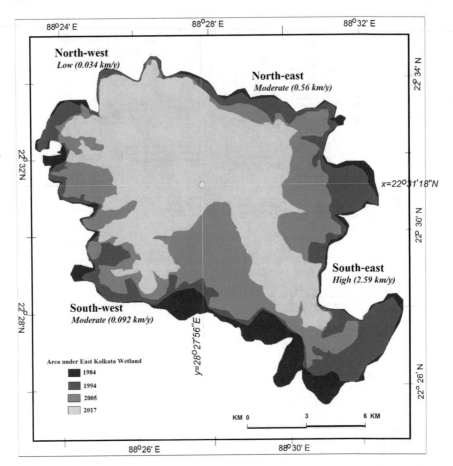

Fig. 13.6 Spatio-temporal variation in the water environment of EKW with rate of changes (assessment was done according to axial extension within EKW). *Source* Prepared by the author based on different earth observatory outputs, 1984–2017

The south-east segment was once linked to the Adi Ganga and palaeo-channels of the Bidyadhari River. Hence, seasonal floods were a regular phenomenon during the monsoon season. But due to the decay of the Bidyadhari channel, this area has a long history of rampant agricultural practices without the use of sewage water. Since the mid-1980s, partial landfilling and conversion of waterbodies for agricultural and orchards practices became popular among locals and led to the proliferation of non-metalled roads. Later, the extension of metalled roads triggered the partial conversion of ponds into rural uses, and the area was occupied by wooded settlements whose residents undertook agricultural practices. Yet the agricultural procedure without a direct link to the sewage recycling activities has prevailed for a long time. The wetland water environment has very little controlling influence on this primary economic activity. Although a few ponds exist, they have very little connections with waste recycling activities in the south-eastern periphery—albeit its northern portion does have a few small ponds with wastewater treatment practices and aquaculture. Actually, this segment is mostly located within the non-core wetland system. Land fragmentation and built-up encroachment were very low during the last three decades; however, owing to the absence of wetland ecosystems, the linear retreat of the water environment in this section of the EKW is high—with a rate of 2.59 km/year due to the lost channel linkage. The south-western section of the EKW experiences rampant land fragmentation, vigorous transformation and built-up encroachment. A large flood area with wetlands of varied sizes was the pivotal characteristic of this area until the late 1980s. However, at present, it is full of infrastructure development—with very little wastewater treatment linkages. Mixed uses of land along with wooded settlements, 'bheries' and agricultural areas characterized the Goalpota, Deara and Kheadaha mouzas, which are located in the central portion of the EKW. The propagation of non-metalled roads has opened new areas to opportunities linked to the construction of residential buildings. Presently, various high-rise enclaves, apartments and gated communities are arising in the south-western periphery. In the compacted wooded settlement areas, isolated ponds with concrete banks and medium to large sized 'bheries' along the canal site were noticed during field observation. The extension of the infrastructure environment along with built-up encroachment is also medium in nature—with a rate of 0.92 km/year.

4 Conclusion

The pace of urban expansion in the light of infrastructural development in the western and northern peripheries welcomes large private farmers within and around existing waterbodies. Their profit-intensive activities increase with the increasing demand for residential and industrial fulfilments—along with governmental initiatives and projects. Presently, the government is trying to operationalize another flyover and expressway in areas of tropical wetland. This situation will add additional risks— which include the declining nature of flora and fauna population, the fragile condition of the carbon sequestration process, loss of water bodies due to the construction of firms, pillars, embankments and roads, fragmentation and re-fragmentation of larger water bodies into small 'bheries' and so on. Subsequently, these 'bheries' will be easily converted into other uses, which may lead to rapid changes of biodiversity and associated ecosystems. Residential, industrial and public utility plots of land create a huge number of pollutants and increase the concentration of heavy metals in the water environment. Encroachment of water bodies due to infrastructural advancement can undermine the distribution of sewage water from the main canals, which increases siltation on the canal beds and thwarts the fishpond by creating small bars in the joining section. The uncontrolled growth of infrastructure and built environments within this sensitive area has converted irregular and non-diametric wetland ecosystems to a regular and iso-diametric structure—which is deeply controlled by urban linkage. The proper demarcation of the EKW boundary, integrated management policy, sustainable land utilization practices, estimation of land carrying capacity, technology-based land use monitoring system and continuous companion along with awareness programme will protect the EKW from fragmentation and sustain its ecology.

Acknowledgements The author is thankful to Mr. S. V. Mishra, Ms. P. Karmakar and Ms. O. Das, research scholar and student in the Dept. of Geography, University of Calcutta, for their continuous support and cartographic assistance.

Appendix

Some glimpses of the land encroachment in the EKW and related issues.

The Infrastructure-driven Changing phenomena in the LULC in EKW

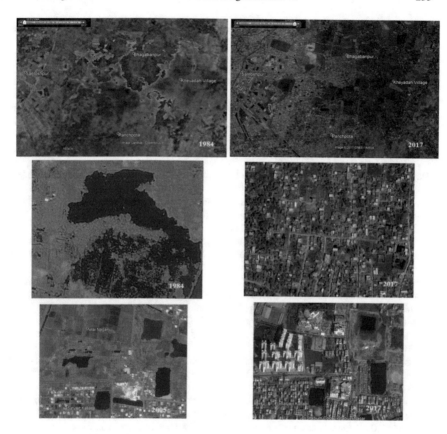

References

Aguilar AG, Ward PM (2003) Globalization, regional development, and megacity expansion in Latin America: analysing Mexico City's peri-urban hinterland. Cities 20:3–21

Alphan H (2003) Land use change and urbanization in Adana, Turkey. Land Degrad Dev 14(6):575–586

Bagchi KG (1944) The Ganga Delta: stages in formation of Ganga Delta. University of Calcutta, pp 50–71

Baker LA, Hope D, Xu Y, Edmonds J, Lauver L (2001) Nitrogen balance for the Central Arizona-Phoenix (CAP) ecosystem. Ecosystems 4(6):582–602

Bhattacharya S, Ganguli A, Bose S, Mukhopadhyay A (2012) Biodiversity, traditional practices and sustainability issue of East Kolkata Wetland: a significance Ramsar site of west Bengal (India). Res Rev Biosci Rev 6(11):340–347

Bose S (2008) Adaptive and integrated management of wastewater and storm water drainage in Kolkata—case study of a Mega City. In: Pahl-Wostl et al (eds) Adaptive and integrated water management: coping with complexity and uncertainty. Springer, New York, pp 341–356

Brown LR (1995) Who will feed China? Wake-up call for a small planet. Worldwatch environmental alert series. W. W. Norton, New York

Chattopadhyaya H (1990) From marsh to township: east of Calcutta, monograph. K. P. Bagchi & Co., Calcutta/New Delhi

Cosentino BJ, Schooley RL (2018) Dispersal and wetland fragmentation. In: Finlayson CM et al (eds) The wetland book. Springer, pp 105–111

Costanza R, d'Arge R, de Groot R et al (1997) The value of the world's ecosystem services and natural capital. Nature 387:253–600

DEC (1945) History of the Ganga Delta, Appendix I-A in report of the Committee to Enquire into the Drainage Conditions of Calcutta and Adjoining Area. Drainage enquiry Committee, Govt. of West Bengal, Calcutta, pp 8–10

Dey D, Banerjee S (2013a) Change in land-use pattern of East Kolkata wetlands: concern and consequences. Urban India 33(1)

Dey D, Banerjee S (2013b) Ecosystem and livelihood support: the story of East Kolkata wetlands. Environ Urban Asia 4(2):325–337

Dewan AM, Yamaguchi Y (2009a) Land use and land cover change in Greater Dhaka, Bangladesh: using remote sensing to promote sustainable urbanization. Environ Monit Assess Appl Geogr. https://doi.org/10.1016/j.apgeog.2008.12.005

Dewan AM, Yamaguchi Y (2009b) Using remote sensing and GIS to detect and monitor land use and land cover change in Dhaka metropolitan of Bangladesh during 1960–2005. Environ Monit Assess 150:237–249

Dewan AM, Yamaguchi Y, Rahaman MZ (2010) Dynamics of land use/cover changes and the analysis of landscape fragmentation in Dhaka Metropolitan Bangladesh. Geo Journal. https://doi.org/10.1007/s10708-010-9399-x (published online 14 Dec 2010)

East Kolkata Wetland Management Authority (2011) Conservation and management plans of east Kolkata wetlands, India: report to be submitted to the Ministry of Environment and Forests, Government of India

Ghosh D (1987) Ecological history of Calcutta's wetland conversion. Environ Conserv 14(3):220

Ghosh D (1999) Wastewater utilization in East Calcutta wetlands. UWEP occasional paper, WASTE, Netherlands

Ghosh D (2005) Ecology and traditional wetland practice, lessons from wastewater utilization in the East Calcutta wetlands. Worldview, Kolkata

Ghosh D (2017). Dispensation of a failed ecologist. The Sunday Statesman, Kolkata, p 07

Ghosh SK, Dutta A (2019) Chandraketugarh theke DumDum (in Bengali language): history of ancient, middle and modern age. Janapath, Kolkata

Grimm NB (2008) Global change and the ecology of cities. Science 319(5864):756–760

Herold M, Clarke KC, Scepan J (2002) Remote sensing and landscape metrics to describe structures and changes in urban land use. Environ Plan Ser A 34:1443–1458

Haque SM, Bandyopadhyay S (2012) Identification of metropolitan core using geo-spatial data for Kolkata, India. Sci Ann Alexandru Ioan Cuza Univ 58(II):185–206

Haque SM, Das A, Ruksana (2019) Ecological footprint reduction instrument. Encyclopaedia of renewable and sustainable materials. Elsevier. https://doi.org/10.1016/b978-0-12-803581-8.11042-2

Hartshorn TA (1992) Interpreting the city: an urban geography, 2nd edn. Wiley, New York

Kilic S, Evrendilek F, Berberoglu S, Demirkesen AC (2006) Environmental monitoring of land-use and land-cover changes in Mediterranean region of Turkey. Environ Monit Assess 114:157–168

Liu J, Linderman M, OUyang Z, An L, Yang J, Zhang F (2002) Ecological degradation in protected areas: the case of Wolong Nature Reserve for giant pandas. Science 292:98–101

McDonnell MJ, Steward TAP, Groffman P, Bohlen P, Pouyat RV, Zipperer WC, Parmelee RW, Carreiro MM, Medley K (1997) Ecosystem processes along an urban-to-rural gradient. Urban Ecosyst 1(1):21–36

Mills G (2007) Cities as agents of global change. Int J Climatol 27:1849–1857

Mukerjee R (1938) Decline of the western delta and estuary since the 17th century. Changing face of Bengal: a study in riverine economy. University of Calcutta, Calcutta, pp 195–199

Nagendra H, Munroe D, Southworth J (2003) From pattern to process: landscape fragmentation and the analysis of land use/land cover change. Agric Ecosyst Environ 101:111–115

Pabi O (2007) Understanding land-use/cover change process for land and environmental resources use management policy in Ghana. Geo-Journal 68:369–383

Pathak CR (2011) The fringe area of Kolkata metropolis: nature, problems and prospects. In: Dikshit JK (ed) The urban fringe of Indian cities. Rawat Publication, Jaipur, pp 113–123

Shaw A (2015) Inner-city and outer-city neighbourhoods in Kolkata: their changing dynamics post liberalization. Environ Urban Asia 6(2):139–153

State Planning Board (1990) Perspective plan for Calcutta: 2011, Govt of West Bengal, pp 197–210

UN HABITAT (2003) The challenges of slum: global report on human settlement 2003. United Nations Human Settlements Programme, Earthscan, London, pp iii–310

Xiao JY et al (2006) Evaluating urban expansion and land use change in Shijiazhuang, China by using GIS and remote sensing. Landsc Urban Plan 75:69–80

Index

© Springer Nature Switzerland AG 2020
S. Bandyopadhyay et al. (eds.), *Water Management in South Asia*,
Contemporary South Asian Studies, https://doi.org/10.1007/978-3-030-35237-0

Printed in the United States
By Bookmasters